Practical Data Analysis

Transform, model, and visualize your data through
hands-on projects, developed in open source tools

Hector Cuesta

BIRMINGHAM - MUMBAI

Practical Data Analysis

First published: October 2013

Production Reference: 1151013

Published by Packt Publishing Ltd.
Livery Place
35 Livery Street
Birmingham B3 2PB, UK.

ISBN 978-1-78328-099-5

www.packtpub.com

Cover Image by Hector Cuesta (hmcuesta.data@gmail.com)

Credits

Author

Hector Cuesta

Reviewers

Dr. Sampath Kumar Kanthala

Mark Kerzner

Ricky J. Sethi, PhD

Dr. Suchita Tripathi

Dr. Jarrell Waggoner

Acquisition Editors

Edward Gordon

Erol Staveley

Lead Technical Editor

Neeshma Ramakrishnan

Technical Editors

Pragnesh Bilimoria

Arwa Manasawala

Manal Pednekar

Project Coordinator

Anugya Khurana

Proofreaders

Jenny Blake

Bridget Braund

Indexer

Hemangini Bari

Graphics

Rounak Dhruv

Abhinash Sahu

Sheetal Aute

Production Coordinator

Arvindkumar Gupta

Cover Work

Arvindkumar Gupta

Foreword

The phrase: From Data to Information, and from Information to Knowledge, has become a cliché but it has never been as fitting as today. With the emergence of Big Data and the need to make sense of the massive amounts of disparate collection of individual datasets, there is a requirement for practitioners of data-driven domains to employ a rich set of analytic methods. Whether during data preparation and cleaning, or data exploration, the use of computational tools has become imperative. However, the complexity of underlying theories represent a challenge for users who wish to apply these methods to exploit the potentially rich contents of available data in their domain. In some domains, text-based data may hold the secret of running a successful business. For others, the analysis of social networks and the classification of sentiments may reveal new strategies for the dissemination of information or the formulation of policy.

My own research and that of my students falls in the domain of computational epidemiology. Designing and implementing tools that facilitate the study of the progression of diseases in a large population is the main focus in this domain. Complex simulation models are expected to predict, or at least suggest, the most likely trajectory of an epidemic. The development of such models depends on the availability or data from which population and disease specific parameters can be extracted. Whether census data, which holds information about the makeup of the population, of medical texts, which describe the progression of disease in individuals, the data exploration represents a challenging task. As many areas that employ data analytics, computational epidemiology is intrinsically multi-disciplinary. While the analysis of some data sources may reveal the number of eggs deposited by a mosquito, other sources may indicate the rate at which mosquitoes are likely to interact with the human population to cause a Dengue and West-Nile Virus epidemic. To convert information to knowledge, computational scientists, biologists, biostatisticians, and public health practitioners must collaborate. It is the availability of sophisticated visualization tools that allows these diverse groups of scientists and practitioners to explore the data and share their insight.

I first met Hector Cuesta during the Fall Semester of 2011, when he joined my Computational Epidemiology Research Laboratory as a visiting scientist. I soon realized that Hector is not just an outstanding programmer, but also a practitioner who can readily apply computational paradigms to problems from different contexts. His expertise in a multitude of computational languages and tools, including Python, CUDA, Hadoop, SQL, and MPI allows him to construct solutions to complex problems from different domains. In this book, Hector Cuesta is demonstrating the application of a variety of data analysis tools on a diverse set of problem domains. Different types of datasets are used to motivate and explore the use of powerful computational methods that are readily applicable to other problem domains. This book serves both as a reference and as tutorial for practitioners to conduct data analysis and move From Data to Information, and from Information to Knowledge.

Armin R. Mikler
Professor of Computer Science and Engineering
Director of the Center for Computational Epidemiology and Response Analysis
University of North Texas

About the Author

Hector Cuesta holds a B.A in Informatics and M.Sc. in Computer Science. He provides consulting services for software engineering and data analysis with experience in a variety of industries including financial services, social networking, e-learning, and human resources.

He is a lecturer in the Department of Computer Science at the Autonomous University of Mexico State (UAEM). His main research interests lie in computational epidemiology, machine learning, computer vision, high-performance computing, big data, simulation, and data visualization.

He helped in the technical review of the books, *Raspberry Pi Networking Cookbook* by *Rick Golden* and *Hadoop Operations and Cluster Management Cookbook* by *Shumin Guo* for Packt Publishing. He is also a columnist at Software Guru magazine and he has published several scientific papers in international journals and conferences. He is an enthusiast of Lego Robotics and Raspberry Pi in his spare time.

You can follow him on Twitter at `https://twitter.com/hmCuesta`.

Acknowledgments

I would like to dedicate this book to my wife Yolanda, my wonderful children Damian and Isaac for all the joy they bring into my life, and to my parents Elena and Miguel for their constant support and love.

I would like to thank my great team at Packt Publishing, particular thanks goes to, Anurag Banerjee, Erol Staveley, Edward Gordon, Anugya Khurana, Neeshma Ramakrishnan, Arwa Manasawala, Manal Pednekar, Pragnesh Bilimoria, and Unnati Shah.

Thanks to my friends, Abel Valle, Oscar Manso, Ivan Cervantes, Agustin Ramos, Dr. Rene Cruz, Dr. Adrian Trueba, and Sergio Ruiz for their helpful suggestions and improvements to my drafts. I would also like to thank the technical reviewers for taking the time to send detailed feedback for the drafts.

I would also like to thank Dr. Armin Mikler for his encouragement and for agreeing to write the foreword of this book. Finally, as an important source of inspiration I would like to mention my mentor and former advisor Dr. Jesus Figueroa-Nazuno.

About the Reviewers

Mark Kerzner holds degrees in Law, Math, and Computer Science. He has been designing software for many years, and Hadoop-based systems since 2008. He is the President of SHMsoft, a provider of Hadoop applications for various verticals, and a co-author of the Hadoop Illuminated book/project. He has authored and co-authored books and patents.

I would like to acknowledge the help of my colleagues, in particular Sujee Maniyam, and last but not least I would acknowledge the help of my multi-talented family.

Dr. Sampath Kumar works as an assistant professor and head of the Department of Applied Statistics at Telangana University. He has completed M.Sc, M.Phl, and Ph.D. in Statistics. He has five years of teaching experience for PG course. He has more than four years of experience in the corporate sector. His expertise is in statistical data analysis using SPSS, SAS, R, Minitab, MATLAB, and so on. He is an advanced programmer in SAS and matlab software. He has teaching experience in different, applied and pure statistics subjects such as forecasting models, applied regression analysis, multivariate data analysis, operations research, and so on for M.Sc students. He is currently supervising Ph.D. scholars.

Ricky J. Sethi is currently the Director of Research for The Madsci Network and a research scientist at University of Massachusetts Medical Center and UMass Amherst. Dr. Sethi's research tends to be interdisciplinary in nature, relying on machine-learning methods and physics-based models to examine issues in computer vision, social computing, and science learning. He received his B.A. in Molecular and Cellular Biology (Neurobiology)/Physics from the University of California, Berkeley, M.S. in Physics/Business (Information Systems) from the University of Southern California, and Ph.D. in Computer Science (Artificial Intelligence/Computer Vision) from the University of California, Riverside. He has authored or co-authored over 30 peer-reviewed papers or book chapters and was also chosen as an NSF Computing Innovation Fellow at both UCLA and USC's Information Sciences Institute.

Dr. Suchita Tripathi did her Ph.D. and M.Sc. at Allahabad University in Anthropology. She also has skills in computer applications and SPSS data analysis software. She has language proficiency in Hindi, English, and Japanese. She learned primary and intermediate level Japanese language from ICAS Japanese language training school, Sendai, Japan and received various certificates. She is the author of six articles and one book. She had two years of teaching experience in the Department of Anthropology and Tribal Development, GGV Central University, Bilaspur (C.G.). Her major areas of research are Urban Anthropology, Anthropology of Disasters, Linguistic and Archeological Anthropology.

I would like to acknowledge my parents and my lovely family for their moral support, and well wishes.

Dr. Jarrell Waggoner is a software engineer at Groupon, working on internal tools to perform sales analytics and demand forecasting. He completed his Ph.D. in Computer Science and Engineering from the University of South Carolina and has worked on numerous projects in the areas of computer vision and image processing, including an NEH-funded document image processing project, a DARPA competition to build an event recognition system, and an interdisciplinary AFOSR-funded materials science image processing project. He is an ardent supporter of free software, having used a variety of open source languages, operating systems, and frameworks in his research. His open source projects and contributions, along with his research work, can be found on GitHub (`https://github.com/malloc47`) and on his website (`http://www.malloc47.com`).

www.PacktPub.com

Support files, eBooks, discount offers and more

You might want to visit www.PacktPub.com for support files and downloads related to your book.

Did you know that Packt offers eBook versions of every book published, with PDF and ePub files available? You can upgrade to the eBook version at www.PacktPub.com and as a print book customer, you are entitled to a discount on the eBook copy. Get in touch with us at service@packtpub.com for more details.

At www.PacktPub.com, you can also read a collection of free technical articles, sign up for a range of free newsletters and receive exclusive discounts and offers on Packt books and eBooks.

http://PacktLib.PacktPub.com

Do you need instant solutions to your IT questions? PacktLib is Packt's online digital book library. Here, you can access, read and search across Packt's entire library of books.

Why Subscribe?

- Fully searchable across every book published by Packt
- Copy and paste, print and bookmark content
- On demand and accessible via web browser

Free Access for Packt account holders

If you have an account with Packt at www.PacktPub.com, you can use this to access PacktLib today and view nine entirely free books. Simply use your login credentials for immediate access.

Table of Contents

Preface

Practical Data Analysis provides a series of practical projects in order to turn data into insight. It covers a wide range of data analysis tools and algorithms for classification, clustering, visualization, simulation, and forecasting. The goal of this book is to help you understand your data to find patterns, trends, relationships, and insight.

This book contains practical projects that take advantage of the MongoDB, D3.js, and Python language and its ecosystem to present the concepts using code snippets and detailed descriptions.

What this book covers

Chapter 1, Getting Started, discusses the principles of data analysis and the data analysis process.

Chapter 2, Working with Data, explains how to scrub and prepare your data for the analysis and also introduces the use of OpenRefine which is a data cleansing tool.

Chapter 3, Data Visualization, shows how to visualize different kinds of data using D3.js, which is a JavaScript Visualization Framework.

Chapter 4, Text Classification, introduces the binary classification using a Naïve Bayes algorithm to classify spam.

Chapter 5, Similarity-based Image Retrieval, presents a project to find the similarity between images using a dynamic time warping approach.

Chapter 6, Simulation of Stock Prices, explains how to simulate stock prices using Random Walk algorithm, visualized with a D3.js animation.

Chapter 7, Predicting Gold Prices, introduces how Kernel Ridge Regression works and how to use it to predict the gold price using time series.

Chapter 8, Working with Support Vector Machines, describes how to use support vector machines as a classification method.

Chapter 9, Modeling Infectious Disease with Cellular Automata, introduces the basic concepts of computational epidemiology simulation and explains how to implement a cellular automaton to simulate an epidemic outbreak using D3.js and JavaScript.

Chapter 10, Working with Social Graphs, explains how to obtain and visualize your social media graph from Facebook using Gephi.

Chapter 11, Sentiment Analysis of Twitter Data, explains how to use the Twitter API to retrieve data from Twitter. We also see how to improve the text classification to perform a sentiment analysis using the Naïve Bayes algorithm implemented in the Natural Language Toolkit (NLTK).

Chapter 12, Data Processing and Aggregation with MongoDB, introduces the basic operations in MongoDB as well as methods for grouping, filtering, and aggregation.

Chapter 13, Working with MapReduce, illustrates how to use the MapReduce programming model implemented in MongoDB.

Chapter 14, Online Data Analysis with IPython and Wakari, explains how to use the Wakari platform and introduces the basic use of Pandas and PIL with IPython.

Appendix, Setting Up the Infrastructure, provides detailed information on installation of the software tools used in this book.

What you need for this book

The basic requirements for this book are as follows:

- Python
- OpenRefine
- D3.js
- mlpy
- Natural Language Toolkit (NLTK)
- Gephi
- MongoDB

Who this book is for

This book is for software developers, analysts, and computer scientists who want to implement data analysis and visualization in a practical way. The book is also intended to provide a self-contained set of practical projects in order to get insight about different kinds of data such as, time series, numerical, multidimensional, social media graphs, and texts. You are not required to have previous knowledge about data analysis, but some basic knowledge about statistics and a general understanding of Python programming is essential.

Conventions

In this book, you will find a number of styles of text that distinguish between different kinds of information. Here are some examples of these styles, and an explanation of their meaning.

Code words in text, database table names, folder names, filenames, file extensions, pathnames, dummy URLs, user input, and Twitter handles are shown as follows: "In this case, we will use the integrate method of the SciPy module to solve the ODE."

A block of code is set as follows:

```
beta = 0.003
gamma = 0.1
sigma = 0.1

def SIRS_model(X, t=0):

   r = scipy.array([- beta*X[0]*X[1] + sigma*X[2]
     , beta*X[0]*X[1] - gamma*X[1]
     , gamma*X[1] ] -sigma*X[2])
   return r
```

When we wish to draw your attention to a particular part of a code block, the relevant lines or items are highlighted as follows:

```
[[215  10    0]
 [153  72    0]
 [ 54 171    0]
 [  2 223    0]
 [  0 225    0]
 [  0 178   47]]
```

```
    [  0  72 153]
    [  0   6 219]
    [  0   0 225]
    [ 47   0 178]
    [153   0  72]
    [219   0   6]
    [225   0   0]]
```

Any command-line input or output is written as follows:

```
db.runCommand( { count: TweetWords })
```

New terms and **important words** are shown in bold. Words that you see on the screen, in menus or dialog boxes for example, appear in the text like this: "Next, as we can see in the following screenshot, we will click on the **Map Reduce** option."

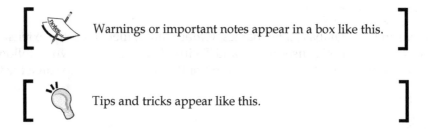

Warnings or important notes appear in a box like this.

Tips and tricks appear like this.

Reader feedback

Feedback from our readers is always welcome. Let us know what you think about this book—what you liked or may have disliked. Reader feedback is important for us to develop titles that you really get the most out of.

To send us general feedback, simply send an e-mail to feedback@packtpub.com, and mention the book title via the subject of your message.

If there is a topic that you have expertise in and you are interested in either writing or contributing to a book, see our author guide on www.packtpub.com/authors.

Customer support

Now that you are the proud owner of a Packt book, we have a number of things to help you to get the most from your purchase.

Downloading the example code

You can download the example code files for all Packt books you have purchased from your account at `http://www.packtpub.com`. If you purchased this book elsewhere, you can visit `http://www.packtpub.com/support` and register to have the files e-mailed directly to you.

Errata

Although we have taken every care to ensure the accuracy of our content, mistakes do happen. If you find a mistake in one of our books—maybe a mistake in the text or the code—we would be grateful if you would report this to us. By doing so, you can save other readers from frustration and help us improve subsequent versions of this book. If you find any errata, please report them by visiting `http://www.packtpub.com/submit-errata`, selecting your book, clicking on the **errata submission form** link, and entering the details of your errata. Once your errata are verified, your submission will be accepted and the errata will be uploaded on our website, or added to any list of existing errata, under the Errata section of that title. Any existing errata can be viewed by selecting your title from `http://www.packtpub.com/support`.

Piracy

Piracy of copyright material on the Internet is an ongoing problem across all media. At Packt, we take the protection of our copyright and licenses very seriously. If you come across any illegal copies of our works, in any form, on the Internet, please provide us with the location address or website name immediately so that we can pursue a remedy.

Please contact us at `copyright@packtpub.com` with a link to the suspected pirated material.

We appreciate your help in protecting our authors, and our ability to bring you valuable content.

Questions

You can contact us at `questions@packtpub.com` if you are having a problem with any aspect of the book, and we will do our best to address it.

1
Getting Started

Data analysis is the process in which raw data is ordered and organized, to be used in methods that help to explain the past and predict the future. Data analysis is not about the numbers, it is about making/asking questions, developing explanations, and testing hypotheses. **Data Analysis** is a multidisciplinary field, which combines **Computer Science**, **Artificial Intelligence & Machine Learning**, **Statistics & Mathematics**, and **Knowledge Domain** as shown in the following figure:

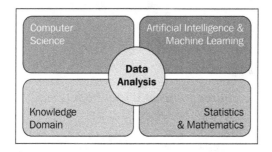

Computer science

Computer science creates the tools for data analysis. The vast amount of data generated has made computational analysis critical and has increased the demand for skills such as programming, database administration, network administration, and high-performance computing. Some programming experience in Python (or any high-level programming language) is needed to understand the chapters.

Artificial intelligence (AI)

According to Stuart Russell and Peter Norvig:

> *"[AI] has to do with smart programs, so let's get on and write some."*

In other words, AI studies the algorithms that can simulate an intelligent behavior. In data analysis, we use AI to perform those activities that require intelligence such as inference, similarity search, or unsupervised classification.

Machine Learning (ML)

Machine learning is the study of computer algorithms to learn how to react in a certain situation or recognize patterns. According to Arthur Samuel (1959),

> *"Machine Learning is a field of study that gives computers the ability to learn without being explicitly programmed."*

ML has a large amount of algorithms generally split in to three groups; given how the algorithm is training:

- Supervised learning
- Unsupervised learning
- Reinforcement learning

Relevant numbers of algorithms are used throughout the book and are combined with practical examples, leading the reader through the process from the data problem to its programming solution.

Statistics

In January 2009, Google's Chief Economist, Hal Varian said,

> *"I keep saying the sexy job in the next ten years will be statisticians. People think I'm joking, but who would've guessed that computer engineers would've been the sexy job of the 1990s?"*

Statistics is the development and application of methods to collect, analyze, and interpret data.

Data analysis encompasses a variety of statistical techniques such as simulation, Bayesian methods, forecasting, regression, time-series analysis, and clustering.

Mathematics

Data analysis makes use of a lot of mathematical techniques such as linear algebra (vector and matrix, factorization, and eigenvalue), numerical methods, and conditional probability in the algorithms. In this book, all the chapters are self-contained and include the necessary math involved.

Knowledge domain

One of the most important activities in data analysis is asking questions, and a good understanding of the knowledge domain can give you the expertise and intuition needed to ask good questions. Data analysis is used in almost all the domains such as finance, administration, business, social media, government, and science.

Data, information, and knowledge

Data are facts of the world. For example, financial transactions, age, temperature, number of steps from my house to my office, are simply numbers. The information appears when we work with those numbers and we can find value and meaning. The information can help us to make informed decisions.

We can talk about knowledge when the data and the information turn into a set of rules to assist the decisions. In fact, we can't store knowledge because it implies theoretical or practical understanding of a subject. However, using predictive analytics, we can simulate an intelligent behavior and provide a good approximation. An example of how to turn data into knowledge is shown in the following figure:

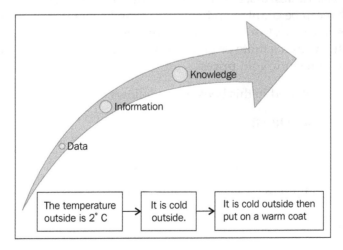

The nature of data

Data is the plural of datum, so it is always treated as plural. We can find data in all the situations of the world around us, in all the structured or unstructured, in continuous or discrete conditions, in weather records, stock market logs, in photo albums, music playlists, or in our Twitter accounts. In fact, data can be seen as the essential raw material of any kind of human activity. According to the *Oxford English Dictionary*:

> *Data are known facts or things used as basis for inference or reckoning.*

As shown in the following figure, we can see **Data** in two distinct ways: **Categorical** and **Numerical**:

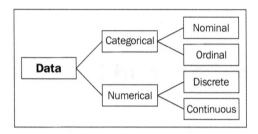

Categorical data are values or observations that can be sorted into groups or categories. There are two types of categorical values, nominal and ordinal. A nominal variable has no intrinsic ordering to its categories. For example, housing is a categorical variable having two categories (own and rent). An ordinal variable has an established ordering. For example, age as a variable with three orderly categories (young, adult, and elder).

Numerical data are values or observations that can be measured. There are two kinds of numerical values, discrete and continuous. Discrete data are values or observations that can be counted and are distinct and separate. For example, number of lines in a code. Continuous data are values or observations that may take on any value within a finite or infinite interval. For example, an economic time series such as historic gold prices.

The kinds of datasets used in this book are as follows:

- E-mails (unstructured, discrete)
- Digital images (unstructured, discrete)
- Stock market logs (structured, continuous)
- Historic gold prices (structured, continuous)

- Credit approval records (structured, discrete)
- Social media friends and relationships (unstructured, discrete)
- Tweets and trending topics (unstructured, continuous)
- Sales records (structured, continuous)

For each of the projects in this book, we try to use a different kind of data. This book is trying to give the reader the ability to address different kinds of data problems.

The data analysis process

When you have a good understanding of a phenomenon, it is possible to make predictions about it. Data analysis helps us to make this possible through exploring the past and creating predictive models.

The data analysis process is composed of the following steps:

- The statement of problem
- Obtain your data
- Clean the data
- Normalize the data
- Transform the data
- Exploratory statistics
- Exploratory visualization
- Predictive modeling
- Validate your model
- Visualize and interpret your results
- Deploy your solution

All these activities can be grouped as shown in the following figure:

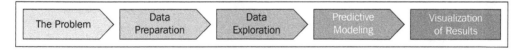

The problem

The problem definition starts with high-level questions such as how to track differences in behavior between groups of customers, or what's going to be the gold price in the next month. Understanding the objectives and requirements from a domain perspective is the key to a successful data analysis project.

Types of data analysis questions are listed as follows:

- Inferential
- Predictive
- Descriptive
- Exploratory
- Causal
- Correlational

Data preparation

Data preparation is about how to obtain, clean, normalize, and transform the data into an optimal dataset, trying to avoid any possible data quality issues such as invalid, ambiguous, out-of-range, or missing values. This process can take a lot of your time. In *Chapter 2, Working with Data*, we go into more detail about working with data, using **OpenRefine** to address the complicated tasks. Analyzing data that has not been carefully prepared can lead you to highly misleading results.

The characteristics of good data are listed as follows:

- Complete
- Coherent
- Unambiguous
- Countable
- Correct
- Standardized
- Non-redundant

Data exploration

Data exploration is essentially looking at the data in a graphical or statistical form trying to find patterns, connections, and relations in the data. Visualization is used to provide overviews in which meaningful patterns may be found.

In *Chapter 3, Data Visualization*, we present a visualization framework (D3.js) and we implement some examples on how to use visualization as a data exploration tool.

Predictive modeling

Predictive modeling is a process used in data analysis to create or choose a statistical model trying to best predict the probability of an outcome. In this book, we use a variety of those models and we can group them in three categories based on its outcome:

	Chapter	Algorithm
Categorical outcome (Classification)	4	Naïve Bayes Classifier
	11	Natural Language Toolkit + Naïve Bayes Classifier
Numerical outcome (Regression)	6	Random Walk
	8	Support Vector Machines
	9	Cellular Automata
	8	Distance Based Approach + k-nearest neighbor
Descriptive modeling (Clustering)	5	**Fast Dynamic Time Warping (FDTW)** + Distance Metrics
	10	Force Layout and Fruchterman-Reingold layout

Another important task we need to accomplish in this step is evaluating the model we chose to be optimal for the particular problem.

The No Free Lunch Theorem proposed by Wolpert in 1996 stated:

> *"No Free Lunch theorems have shown that learning algorithms cannot be universally good."*

The model evaluation helps us to ensure that our analysis is not over-optimistic or over-fitted. In this book, we are going to present two different ways to validate the model:

- **Cross-validation**: We divide the data into subsets of equal size and test the predictive model in order to estimate how it is going to perform in practice. We will implement cross-validation in order to validate the robustness of our model as well as evaluate multiple models to identify the best model based on their performance.
- **Hold-Out**: Mostly, large dataset is randomly divided in to three subsets: training set, validation set, and test set.

Visualization of results

This is the final step in our analysis process and we need to answer the following questions:

How is it going to present the results?

For example, in tabular reports, 2D plots, dashboards, or infographics.

Where is it going to be deployed?

For example, in hard copy printed, poster, mobile devices, desktop interface, or web.

Each choice will depend on the kind of analysis and a particular data. In the following chapters, we will learn how to use standalone plotting in Python with `matplotlib` and web visualization with `D3.js`.

Quantitative versus qualitative data analysis

Quantitative and qualitative analysis can be defined as follows:

- **Quantitative data**: It is numerical measurements expressed in terms of numbers
- **Qualitative data**: It is categorical measurements expressed in terms of natural language descriptions

As shown in the following figure, we can observe the differences between quantitative and qualitative analysis:

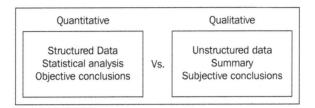

Quantitative analytics involves analysis of numerical data. The type of the analysis will depend on the level of measurement. There are four kinds of measurements:

- **Nominal**: Data has no logical order and is used as classification data

- **Ordinal**: Data has a logical order and differences between values are not constant

- **Interval**: Data is continuous and depends on logical order. The data has standardized differences between values, but does not include zero

- **Ratio**: Data is continuous with logical order as well as regular interval differences between values and may include zero

Qualitative analysis can explore the complexity and meaning of social phenomena. Data for qualitative study may include written texts (for example, documents or email) and/or audible and visual data (for example, digital images or sounds). In *Chapter 11*, *Sentiment Analysis of Twitter Data*, we present a sentiment analysis from Twitter data as an example of qualitative analysis.

Importance of data visualization

The goal of the data visualization is to expose something new about the underlying patterns and relationships contained within the data. The visualization not only needs to look good but also meaningful in order to help organizations make better decisions. Visualization is an easy way to jump into a complex dataset (small or big) to describe and explore the data efficiently.

Many kinds of data visualizations are available such as bar chart, histogram, line chart, pie chart, heat maps, frequency Wordle (as shown in the following figure) and so on, for one variable, two variables, and many variables in one, two, or three dimensions.

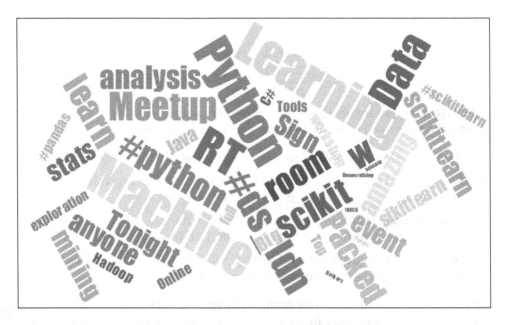

Data visualization is an important part of our data analysis process because it is a fast and easy way to do an exploratory data analysis through summarizing their main characteristics with a visual graph.

The goals of exploratory data analysis are listed as follows:

- Detection of data errors
- Checking of assumptions
- Finding hidden patterns (such as tendency)
- Preliminary selection of appropriate models
- Determining relationships between the variables

We will get into more detail about data visualization and exploratory data analysis in *Chapter 3, Data Visualization*.

What about big data?

Big data is a term used when the data exceeds the processing capacity of typical database. We need a big data analytics when the data grows quickly and we need to uncover hidden patterns, unknown correlations, and other useful information.

There are three main features in big data:

- **Volume**: Large amounts of data
- **Variety**: Different types of structured, unstructured, and multi-structured data
- **Velocity**: Needs to be analyzed quickly

As shown in the following figure, we can see the interaction between the three Vs:

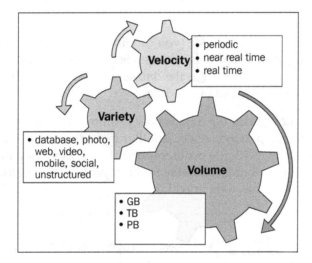

Big data is the opportunity for any company to gain advantages from data aggregation, data exhaust, and metadata. This makes big data a useful business analytic tool, but there is a common misunderstanding about what big data is.

The most common architecture for big data processing is through **MapReduce**, which is a programming model for processing large datasets in parallel using a distributed cluster.

Apache Hadoop is the most popular implementation of MapReduce to solve large-scale distributed data storage, analysis, and retrieval tasks. However, MapReduce is just one of the three classes of technologies for storing and managing big data. The other two classes are **NoSQL** and **massively parallel processing** (**MPP**) data stores. In this book, we implement MapReduce functions and NoSQL storage through **MongoDB**, see *Chapter 12, Data Processing and Aggregation with MongoDB* and *Chapter 13, Working with MapReduce*.

MongoDB provides us with document-oriented storage, high availability, and map/reduce flexible aggregation for data processing.

A paper published by the IEEE in 2009, *The Unreasonable Effectiveness of Data* states:

> *But invariably, simple models and a lot of data trump over more elaborate models based on less data.*

This is a fundamental idea in big data (you can find the full paper at http://bit.ly/1dvHCom). The trouble with real world data is that the probability of finding false correlations is high and gets higher as the datasets grow. That's why, in this book, we focus on meaningful data instead of big data.

One of the main challenges for big data is how to store, protect, backup, organize, and catalog the data in a petabyte scale. Another main challenge of big data is the concept of data ubiquity. With the proliferation of smart devices with several sensors and cameras the amount of data available for each person increases every minute. Big data must process all this data in real time.

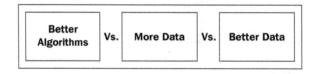

Sensors and cameras

Interaction with the outside world is highly important in data analysis. Using sensors such as **RFID** (**Radio-frequency identification**) or a smartphone to scan a **QR code** (**Quick Response Code**) is an easy way to interact directly with the customer, make recommendations, and analyze consumer trends.

On the other hand, people are using their smartphones all the time, using their cameras as a tool. In *Chapter 5, Similarity-based Image Retrieval*, we will use these digital images to perform search by image. This can be used, for example, in face recognition or to find reviews of a restaurant just by taking a picture of the front door.

The interaction with the real world can give you a competitive advantage and a real-time data source directly from the customer.

Social networks analysis

Formally, the **SNA** (**social network analysis**) performs the analysis of social relationships in terms of network theory, with nodes representing individuals and ties representing relationships between the individuals, as we can see in the following figure. The social network creates groups of related individuals (friendship) based on different aspects of their interaction. We can find important information such as hobbies (for product recommendation) or who has the most influential opinion in the group (centrality). We will present in *Chapter 10, Working with Social Graphs,* a project; who is your closest friend and we'll show a solution for Twitter clustering.

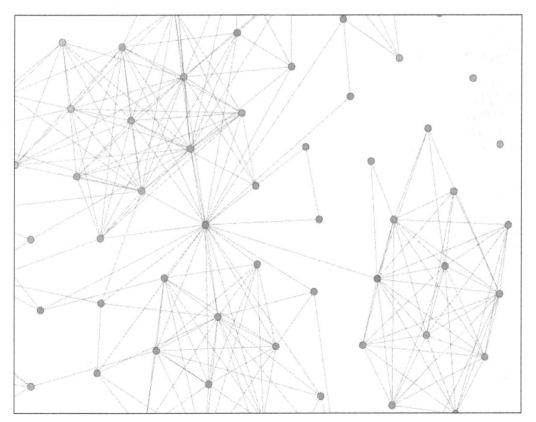

Social networks are strongly connected and these connections are often not symmetric. This makes the SNA computationally expensive, and needs to be addressed with high-performance solutions that are less statistical and more algorithmic.

The visualization of a social network can help us to get a good insight into how people are connected. The exploration of the graph is done through displaying nodes and ties in various colors, sizes, and distributions. The D3.js library has animation capabilities that enable us to visualize the social graph with an interactive animation. These help us to simulate behaviors such as information diffusion or distance between nodes.

Facebook processes more than 500 TB data daily (images, text, video, likes, and relationships), this amount of data needs non-conventional treatment such as NoSQL databases and MapReduce frameworks, in this book, we work with MongoDB—a document-based NoSQL database, which also has great functions for aggregations and MapReduce processing.

Tools and toys for this book

The main goal of this book is to provide the reader with self-contained projects ready to deploy, in order to do this, as you go through the book you will use and implement tools such as Python, D3, and MongoDB. These tools will help you to program and deploy the projects. You also can download all the code from the author's GitHub repository https://github.com/hmcuesta.

You can see a detailed installation and setup process of all the tools in *Appendix, Setting Up the Infrastructure*.

Why Python?

Python is a scripting language—an interpreted language with its own built-in memory management and good facilities for calling and cooperating with other programs. There are two popular Versions, 2.7 or 3.x, in this book, we will focused on the 3.x Version because it is under active development and has already seen over two years of stable releases.

Python is multi-platform, which runs on Windows, Linux/Unix, and Mac OS X, and has been ported to the Java and .NET virtual machines. Python has powerful standard libraries and a wealth of third-party packages for numerical computation and machine learning such as NumPy, SciPy, pandas, SciKit, mlpy, and so on.

Python is excellent for beginners, yet great for experts and is highly scalable — suitable for large projects as well as small ones. Also it is easily extensible and object-oriented.

Python is widely used by organizations such as Google, Yahoo Maps, NASA, RedHat, Raspberry Pi, IBM, and so on.

A list of organizations using Python is available at `http://wiki.python.org/moin/OrganizationsUsingPython`.

Python has excellent documentation and examples at `http://docs.python.org/3/`.

Python is free to use, even for commercial products, download is available for free from `http://python.org/`.

Why mlpy?

mlpy (**Machine Learning Python**) is a Python module built on top of `NumPy`, `SciPy`, and the GNU Scientific Libraries. It is open source and supports Python 3.x. The `mlpy` module has a large amount of machine learning algorithms for supervised and unsupervised problems.

Some of the features of `mlpy` that will be used in this book are as follows:

- We will perform a numeric regression with **kernel ridge regression (KRR)**
- We will explore the dimensionality reduction through **principal component analysis (PCA)**
- We will work with **support vector machines (SVM)** for classification
- We will perform text classification with Naive Bayes
- We will see how different two time series are with **dynamic time warping (DTW)** distance metric

We can download the latest Version of `mlpy` from `http://mlpy.sourceforge.net/`.

For reference you can refer to the paper mply: Machine Learning Python (`http://arxiv.org/abs/1202.6548`) submitted in 2012 by D. Albanese, R. Visintainer, S. Merler, S. Riccadonna, G. Jurman, and C. Furlanello.

Why D3.js?

D3.js (**Data-Driven Documents**) was developed by Mike Bostock. D3 is a JavaScript library for visualizing data and manipulating the document object model that runs in a browser without a plugin. In `D3.js` you can manipulate all the elements of the **DOM** (**Document Object Model**); it is as flexible as the client-side web technology stack (HTML, CSS, and SVG).

`D3.js` supports large datasets and includes animation capabilities that make it a really good choice for web visualization.

D3 has an excellent documentation, examples, and community at `https://github.com/mbostock/d3/wiki/Gallery` and `https://github.com/mbostock/d3/wiki`.

You can download the latest Version of `D3.js` from `http://d3js.org/d3.v3.zip`.

Why MongoDB?

NoSQL (**Not only SQL**) is a term that covers different types of data storage technologies, used when you can't fit your business model into a classical relational data model. NoSQL is mainly used in Web 2.0 and in social media applications.

MongoDB is a document-based database. This means that MongoDB stores and organizes the data as a collection of documents that gives you the possibility to store the view models almost exactly like you model them in the application. Also, you can perform complex searches for data and elementary data mining with MapReduce.

MongoDB is highly scalable, robust, and perfect to work with JavaScript-based web applications because you can store your data in a **JSON** (**JavaScript Object Notation**) document and implement a flexible schema which makes it perfect for no structured data.

MongoDB is used by highly recognized corporations such as Foursquare, Craigslist, Firebase, SAP, and Forbes. We can see a detailed list at `http://www.mongodb.org/about/production-deployments/`.

MongoDB has a big and active community and well-written documentation at `http://docs.mongodb.org/manual/`.

MongoDB is easy to learn and it's free, we can download MongoDB from `http://www.mongodb.org/downloads`.

Summary

In this chapter, we presented an overview of the data analysis ecosystem, explaining basic concepts of the data analysis process, tools, and some insight into the practical applications of the data analysis. We have also provided an overview of the different kinds of data; numerical and categorical. We got into the nature of data, structured (databases, logs, and reports) and unstructured (image collections, social networks, and text mining). Then, we introduced the importance of data visualization and how a fine visualization can help us in the exploratory data analysis. Finally we explored some of the concepts of big data and social networks analysis.

In the next chapter, we will work with data, cleaning, processing, and transforming, using Python and OpenRefine.

Downloading the example code

You can download the example code files for all Packt books you have purchased from your account at `http://www.packtpub.com`. If you purchased this book elsewhere, you can visit `http://www.packtpub.com/support` and register to have the files e-mailed directly to you.

2
Working with Data

Building real world's data analytics requires accurate data. In this chapter we discuss how to obtain, clean, normalize, and transform raw data into a standard format such as **Comma-Separated Values (CSV)** or **JavaScript Object Notation (JSON)** using OpenRefine.

In this chapter we will cover:

- Datasource
    - Open data
    - Text files
    - Excel files
    - SQL databases
    - NoSQL databases
    - Multimedia
    - Web scraping

- Data scrubbing
    - Statistical methods
    - Text parsing
    - Data transformation

- Data formats
    - CSV
    - JSON
    - XML
    - YAML

- Getting started with OpenRefine

Datasource

Datasource is a term used for all the technology related to the extraction and storage of data. A datasource can be anything from a simple text file to a big database. The raw data can come from observation logs, sensors, transactions, or user's behavior.

In this section we will take a look into the most common forms for datasource and datasets.

A dataset is a collection of data, usually presented in tabular form. Each column represents a particular variable, and each row corresponds to a given member of the data, as is shown in the following figure:

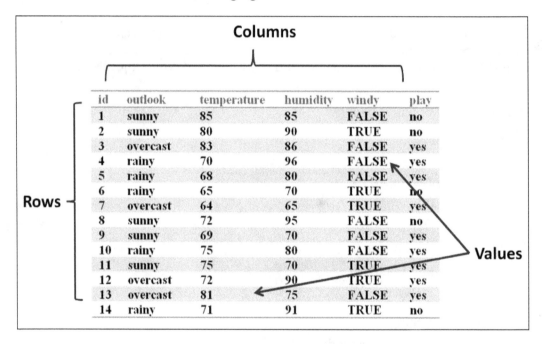

A dataset represents a physical implementation of a datasource; the common features of a dataset are as follows:

- Dataset characteristics (such as multivariate or univariate)
- Number of instances
- Area (for example life, business, and so on)
- Attribute characteristics (namely, real, categorical, and nominal)

- Number of attributes
- Associated tasks (such as classification or clustering)
- Missing Values

Open data

Open data is data that can be used, re-use, and redistributed freely by anyone for any purpose. Following is a short list of repositories and databases for open data:

- Datahub is available at `http://datahub.io/`
- Book-Crossing Dataset is available at `http://www.informatik.uni-freiburg.de/~cziegler/BX/`
- World Health Organization is available at `http://www.who.int/research/en/`
- The World Bank is available at `http://data.worldbank.org/`
- NASA is available at `http://data.nasa.gov/`
- United States Government is available at `http://www.data.gov/`
- Machine Learning Datasets is available at `http://bitly.com/bundles/bigmlcom/2`
- Scientific Data from University of Muenster is available at `http://data.uni-muenster.de/`
- Hilary Mason research-quality datasets is available at `https://bitly.com/bundles/hmason/1`

> Other interesting sources of data come from the data mining and knowledge discovery competitions such as ACM-KDD Cup or Kaggle platform, in most cases the datasets are still available, even after the competition is closed.
>
> Check out the ACM-KDD Cup at the link `http://www.sigkdd.org/kddcup/index.php`.
>
> And Kaggle available at `http://www.kaggle.com/competitions`.

Text files

The text files are commonly used for storage of data, because it is easy to transform into different formats, and it is often easier to recover and continue processing the remaining contents than with other formats. Large amounts of data come in text format from logs, sensors, e-mails, and transactions. There are several formats for text files such as CSV (comma delimited), TSV (tab delimited), **Extensible Markup Language** (**XML**) and (JSON) (see the *Data formats* section).

Excel files

MS-Excel is probably the most used and also the most underrated data analysis tool. In fact Excel has some good points such as filtering, aggregation functions, and using Visual Basis for Application you can make **Structured Query Language (SQL)** – such as queries with the sheets or with an external database.

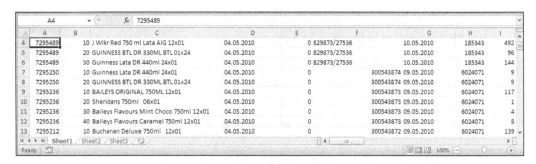

Excel provides us with some visualization tools and we can extend the analysis capabilities of Excel (Version 2010) by installing the Analysis ToolPak that includes functions for Regression, Correlation, Covariance, Fourier Analysis, and so on. For more information about the Analysis ToolPak check the link http://bit.ly/ZQKwSa.

Some Excel disadvantages are that missing values are handled inconsistently and there is no record of how an analysis was accomplished. In the case of the Analysis ToolPak, it can only work with one sheet at a time. That's why Excel is a poor choice for statistical analysis beyond the basic examples.

We can easily transform Excel files (.xls) into another text file format such as CSV, TSV, or even XML. To export the Excel sheet just go to **File** menu, select the option **Save & Send**, and in **Change File Type** select your preferred format such as **CSV (Comma delimited)**.

SQL databases

A database is an organized collection of data. SQL is a database language for managing and manipulating data in **Relational Database Management Systems (RDBMS)**. The **Database Management Systems (DBMS)** are responsible for maintaining the integrity and security of stored data, and for recovering information if the system fails. SQL Language is split into two subsets of instructions, the **Data Definition Language (DDL)** and **Data Manipulation Language (DML)**.

The data is organized in schemas (database) and divided into tables related by logical relationships, where we can retrieve the data by making queries to the main schema, as is shown in the following screenshot:

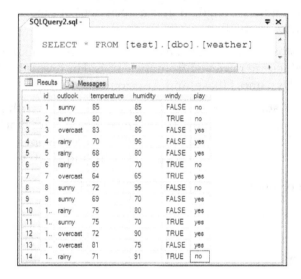

DDL allows us to create, delete, and alter database tables. We can also define keys to specify relationships between tables, and implement constraints between database tables.

- `CREATE TABLE`: This command creates a new table
- `ALTER TABLE`: This command alters a table
- `DROP TABLE`: This command deletes a table

DML is a language which enables users to access and manipulate data.

- `SELECT`: This command retrieves data from the database
- `INSERT INTO`: This command inserts new data into the database
- `UPDATE`: This command modifies data in the database
- `DELETE`: This command deletes data in the database

NoSQL databases

Not only SQL (NoSQL) is a term used in several technologies where the nature of the data does not require a relational model. NoSQL technologies allow working with a huge quantity of data, higher availability, scalability, and performance.

See *Chapter 12, Data Processing and Aggregation with MongoDB* and *Chapter 13, Working with MapReduce,* for extended examples of document store database MongoDB.

The most common types of NoSQL data stores are:

- **Document store**: Data is stored and organized as a collection of documents. The model schema is flexible and each collection can handle any number of fields. For example, MongoDB uses a document of type BSON (binary format of JSON) and CouchDB uses a JSON document.

- **Key-value store**: Data is stored as key-value pairs without a predefined schema. Values are retrieved from their keys. For example, Apache Cassandra, Dynamo, HBase, and Amazon SimpleDB.

- **Graph-based store**: Data is stored in graph structures with nodes, edges, and properties using the computer science graph theory for storing and retrieving data. These kinds of databases are excellent to represent social network relationships. For example, Neo4js, InfoGrid, and Horton.

For more information about NoSQL see the following link:

```
http://nosql-database.org/
```

Multimedia

The increasing number of mobile devices makes it a priority of data analysis to acquire the ability to extract semantic information from multimedia datasources. Datasources include directly perceivable media such as audio, image, and video. Some of the applications for these kinds of datasources are as follows:

- Content-based image retrieval
- Content-based video retrieval
- Movie and video classification
- Face recognition
- Speech recognition
- Audio and music classification

In *Chapter 5, Similarity-based Image Retrieval,* we present a similarity-based image search engine using Caltech256 that is an image dataset with over 30,600 images.

Web scraping

When we want to obtain data, a good place to start is in the web. Web scraping refers to an application that processes the HTML of a web page to extract data for manipulation. Web scraping applications will simulate a person viewing a website with a browser. In the following example, we assume we want to get the current gold price from the website www.gold.org, as is shown in the following screenshot:

Then we need to inspect the **Gold Spot Price** element in the website, where we will find the following HTML tag:

```
<td class="value" id="spotpriceCellAsk">1,573.85</td>
```

We can observe an id, spotpriceCellAsk in the td tag; this is the element we will get with the next Python code.

For this example, we will use the library BeautifulSoup Version 4, in Linux we can install it from the system package manager, we need to open a Terminal and execute the next command:

```
$ apt-get install python-bs4
```

For windows we need to download the library from the following link:

http://crummy.com/software/BeautifulSoup/bs4/download/

To install it, just execute in the command line:

```
$ python setup.py install
```

1. First we need to import the libraries BeautifulSoup and urllib.request

```
from bs4 import BeautifulSoup
import urllib.request
from time import sleep
from datetime import datetime
```

2. Then we use the function `getGoldPrice` to retrieve the current price from the website, in order to do this we need to provide the URL to make the request and read the entire page.

```
req = urllib.request.urlopen(url)
page = req.read()
```

3. Next, we use `BeautifulSoup` to parse the page (creating a list of all the elements of the page) and ask for the element `td` with the `id`, `spotpriceCellAsk`:

```
scraping = BeautifulSoup(page)
price= scraping.findAll("td",attrs={"id":"spotpriceCellAsk"})[0].
text
```

4. Now we return the variable `price` with the current gold price, this value changes every minute on the website, in this case, we want all the values in an hour, so we call the function `getGoldPrice` in a for loop 60 times, making the script wait 59 seconds between each call.

```
for x in range(0,60):
...
   sleep(59)
```

5. Finally, we save the result in a file `goldPrice.out` and include the current date time in the format HH:MM:SS (A.M. or P.M.), for example, 11:35:42PM, separated by a comma.

```
with open("goldPrice.out","w") as f:
...
        sNow = datetime.now().strftime("%I:%M:%S%p")
        f.write("{0}, {1} \n ".format(sNow, getGoldPrice()))
```

The function `datetime.now().strftime` creates a string representing the time under the control of an explicit format string `"%I:%M:%S%p"`, where `%I` represents hour as decimal number from 0 to 12, `%M` represents minute as a decimal number from 00 to 59, `%S` represents second as a decimal number from 00 to 61, and `%p` represent either A.M. or P.M.

A list of complete format directives can be found on the following link:

```
http://docs.python.org/3.2/library/datetime.html
```

The following is the full script:

```python
from bs4 import BeautifulSoup
import urllib.request
from time import sleep
from datetime import datetime
def getGoldPrice():
    url = "http://gold.org"
    req = urllib.request.urlopen(url)
    page = req.read()
    scraping = BeautifulSoup(page)
    price= scraping.findAll("td",attrs={"id":"spotpriceCellAsk"})[0]
    .text
    return price

with open("goldPrice.out","w") as f:
    for x in range(0,60):
        sNow = datetime.now().strftime("%I:%M:%S%p")
        f.write("{0}, {1} \n ".format(sNow, getGoldPrice()))
        sleep(59)
```

 You can download the full script (WebScraping.py) from the author's GitHub repository, which is available at https://github.com/hmcuesta/PDA_Book/tree/master/Chapter2

The output file, goldPrice.out, will look as follows:

```
11:35:02AM, 1481.25
11:36:03AM, 1481.26
11:37:02AM, 1481.28
11:38:04AM, 1481.25
11:39:03AM, 1481.22

...
```

Data scrubbing

Data scrubbing, also called data cleansing, is the process of correcting or removing data in a dataset that is incorrect, inaccurate, incomplete, improperly formatted, or duplicated.

The result of the data analysis process not only depends on the algorithms, it also depends on the quality of the data. That's why the next step after obtaining the data, is data scrubbing. In order to avoid dirty data our dataset should possess the following characteristics:

- Correct
- Completeness
- Accuracy
- Consistency
- Uniformity

The dirty data can be detected by applying some simple statistical data validation also by parsing the texts or deleting duplicate values. Missing or sparse data can lead you to highly misleading results.

Statistical methods

In this method we need some context about the problem (knowledge domain) to find values that are unexpected and thus erroneous, even if the data type match but the values are out of the range, it can be resolved by setting the values to an average or mean value. Statistical validations can be used to handle missing values which can be replaced by one or more probable values using Interpolation or by reducing the dataset using Decimation.

- **Mean**: This is the value calculated by summing up all values and then dividing by the number of values.
- **Median**: The median is the middle value in a sorted list of values.
- **Range Constraints**: The numbers or dates should fall within a certain range. That is, they have minimum and/or maximum possible values.
- **Clustering**: Usually, when we obtain data directly from the user some values include ambiguity or refer to the same value with a typo. For example, Buchanan Deluxe 750ml 12 x 01 and Buchanan Deluxe 750ml 12 x 01., which are different only by a dot, or in the case of Microsoft or MS instead of Microsoft Corporation which refer to the same company and all values are valid. In those cases, grouping can help us to get accurate data and eliminate the duplicated ones, enabling a faster identification of unique values.

Text parsing

We perform parsing to help us to validate if a string of data is well formatted and avoid syntax errors.

Regular expression patterns, usually text fields, would have to be validated in this way. For example, dates, e-mails, phone numbers, and IP addresses. Regex is a common abbreviation for regular expression.

In Python we will use the re module to implement regular expressions. We can perform text search and pattern validations.

Firstly, we need to import the re module.

```
import re
```

In the following examples, we will implement three of the most common validations (e-mail, IP address, and date format):

- E-mail validation:

```
myString = 'From: readers@packt.com (readers email)'
result = re.search('([\w.-]+)@([\w.-]+)', myString)
if result:
    print (result.group(0))
    print (result.group(1))
    print (result.group(2))
```
Output:
```
>>> readers@packt.com
>>> readers
>>> packt.com
```

The function search() scans through a string, searching for any location where the regex might match. The function group() helps us to return the string matched by the regex. The pattern \w matches any alphanumeric character and is equivalent to the class (a-z, A-Z, 0-9_).

- IP address validation:

```
isIP = re.compile('\d{1,3}\.\d{1,3}\.\d{1,3}\.\d{1,3}')
myString = " Your IP is:  192.168.1.254  "
result = re.findall(isIP,myString)
print(result)
```
Output:
```
>>> 192.168.1.254
```

The function `findall()` finds all the substrings where the regex matches, and returns them as a list. The pattern `\d` matches any decimal digit, is equivalent to the class `[0-9]`.

- Date format:

```
myString = "01/04/2001"
isDate = re.match('[0-1][0-9]\/[0-3][0-9]\/[1-2][0-9]{3}',
    myString)
if isDate:
    print("valid")
else:
    print("invalid")
Output:
>>> 'valid'
```

The function match() finds if the regex matches with the string. The pattern implements the class `[0-9]` in order to parse the date format.

> For more information about regular expressions, visit the link
> http://docs.python.org/3.2/howto/regex.html#regex-howto.

Data transformation

Data transformation is usually related to databases and data warehouses, where values from a source format are extracted, transformed, and loaded in a destination format.

Extract, Transform, and Load (ETL) obtains data from datasources, performs some transformation function depending on our data model and loads the result data into destination.

- Data extraction allows us to obtain data from multiple datasources, such as relational databases, data streaming, text files (JSON, CSV, XML), and NoSQL databases.

- Data transformation allows us to cleanse, convert, aggregate, merge, replace, validate, format, and split data.

- Data loading allows us to load data into destination format, such as relational databases, text files (JSON, CSV, XML), and NoSQL databases.

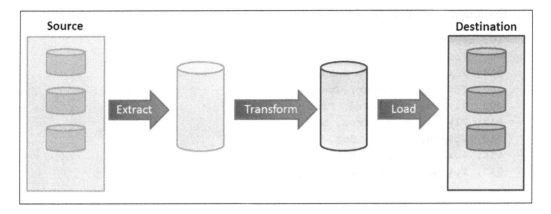

 In statistics, data transformation refers to the application of a mathematical function to the dataset or time series points.

Data formats

When we are working with data for human consumption the easiest way to store it is through text files. In this section, we will present parsing examples of the most common formats such as CSV, JSON, and XML. These examples will be very helpful in the next chapters.

 The dataset used for these examples is a list of Pokémon characters by National Pokedex number, obtained at the URL `http://bulbapedia. bulbagarden.net/`.

All the scripts and dataset files can be found in the author's GitHub repository available at the URL `https://github.com/hmcuesta/ PDA_Book/tree/master/Chapter3/`.

CSV

CSV is a very simple and common open format for table, such as data, which can be exported and imported by most of the data analysis tools. CSV is a plain text format this means that the file is a sequence of characters, with no data that has to be interpreted instead, for example, binary numbers.

There are many ways to parse a CSV file from Python, and in a moment we will discuss two of them:

The first eight records of the CSV file (`pokemon.csv`) look as follows:

```
id, typeTwo, name, type
001, Poison, Bulbasaur, Grass
002, Poison, Ivysaur, Grass
003, Poison, Venusaur, Grass
006, Flying, Charizard, Fire
012, Flying, Butterfree, Bug
013, Poison, Weedle, Bug
014, Poison, Kakuna, Bug
015, Poison, Beedrill, Bug
. . .
```

Parsing a CSV file with the csv module

Firstly, we need to import the csv module:

```
import csv
```

Then we open the file `.csv` and with the function `csv.reader(f)` we parse the file:

```
with open("pokemon.csv") as f:
    data = csv.reader(f)
    #Now we just iterate over the reader

    for line in data:
        print(" id: {0} , typeTwo: {1}, name:  {2}, type: {3}"
              .format(line[0],line[1],line[2],line[3]))
```

```
Output:
[(1, b' Poison', b' Bulbasaur', b' Grass')
 (2, b' Poison', b' Ivysaur', b' Grass')
 (3, b' Poison', b' Venusaur', b' Grass')
 (6, b' Flying', b' Charizard', b' Fire')
    (12, b' Flying', b' Butterfree', b' Bug')
    . . .]
```

Parsing a CSV file using NumPy

Perform the following steps for parsing a CSV file:

1. Firstly, we need to import the `numpy` library:

    ```
    import numpy as np
    ```

2. NumPy provides us with the `genfromtxt` function, which receives four parameters. First, we need to provide the name of the file `pokemon.csv`. Then we skip first line as a header (`skip_header`). Next we need to specify the data type (`dtype`). Finally, we will define the comma as the `delimiter`.

    ```
    data = np.genfromtxt("pokemon.csv"
                        ,skip_header=1
                        ,dtype=None
                        ,delimiter=',')
    ```

3. Then just print the result.

    ```
    print(data)
    ```

    ```
    Output:
    id: id , typeTwo: typeTwo, name:  name, type: type
    id:  001 , typeTwo:  Poison, name:   Bulbasaur, type:  Grass
    id:  002 , typeTwo:  Poison, name:   Ivysaur, type:  Grass
    id:  003 , typeTwo:  Poison, name:   Venusaur, type:  Grass
    id:  006 , typeTwo:  Flying, name:   Charizard, type:  Fire
    . . .
    ```

JSON

JSON is a common format to exchange data. Although it is derived from JavaScript, Python provides us with a library to parse JSON.

Parsing a JSON file using json module

The first three records of the JSON file (`pokemon.json`) look as follows:

```
[
    {
        "id": " 001",
        "typeTwo": " Poison",
        "name": " Bulbasaur",
        "type": " Grass"
    },
    {
```

```
            "id": " 002",
            "typeTwo": " Poison",
            "name": " Ivysaur",
            "type": " Grass"
        },
        {

            "id": " 003",
            "typeTwo": " Poison",
            "name": " Venusaur",
            "type": " Grass"
        },
    . . .]
```

Firstly, we need to import the json module and pprint (pretty-print) module.

```
import json
from pprint import pprint
```

Then we open the file `pokemon.json` and with the function `json.loads` we parse the file.

```
with open("pokemon.json") as f:
    data = json.loads(f.read())
```

Finally, just print the result with the function `pprint`.

```
pprint(data)
```

```
Output:
```

```
[{'id': ' 001', 'name': ' Bulbasaur', 'type': ' Grass', 'typeTwo': '
Poison'},
 {'id': ' 002', 'name': ' Ivysaur', 'type': ' Grass', 'typeTwo': '
Poison'},
 {'id': ' 003', 'name': ' Venusaur', 'type': ' Grass', 'typeTwo': '
Poison'},
 {'id': ' 006', 'name': ' Charizard', 'type': ' Fire', 'typeTwo': '
Flying'},
 {'id': ' 012', 'name': ' Butterfree', 'type': ' Bug', 'typeTwo': '
Flying'}, . . . ]
```

XML

According with to World Wide Web Consortium (W3C) available at
`http://www.w3.org/XML/`

> *Extensible Markup Language (XML) is a simple, very flexible text format derived from SGML (ISO 8879). Originally designed to meet the challenges of large-scale electronic publishing, XML is also playing an increasingly important role in the exchange of a wide variety of data on the Web and elsewhere.*

The first three records of the XML file (`pokemon.xml`) look as follows:

```
<?xml version="1.0" encoding="UTF-8" ?>
<pokemon>
  <row>
    <id> 001</id>
    <typeTwo> Poison</typeTwo>
    <name> Bulbasaur</name>
    <type> Grass</type>
  </row>
  <row>
    <id> 002</id>
    <typeTwo> Poison</typeTwo>
    <name> Ivysaur</name>
    <type> Grass</type>
  </row>
  <row>
    <id> 003</id>
    <typeTwo> Poison</typeTwo>
    <name> Venusaur</name>
    <type> Grass</type>
  </row>
  . . .
</pokemon>
```

Parsing an XML file in Python using xml module

Firstly, we need to import the `ElementTree` object from xml module.

```
from xml.etree import ElementTree
```

Then we open the file `"pokemon.xml"` and with the function `ElementTree.parse` we parse the file.

```
with open("pokemon.xml") as f:
    doc = ElementTree.parse(f)
```

Finally, just print each `'row'` element with the `findall` function:

```
for node in doc.findall('row'):
    print("")
    print("id: {0}".format(node.find('id').text))
    print("typeTwo: {0}".format(node.find('typeTwo').text))
    print("name: {0}".format(node.find('name').text))
    print("type: {0}".format(node.find('type').text))
```

```
Output:

id:   001
typeTwo:   Poison
name:   Bulbasaur
type:   Grass

id:   002
typeTwo:   Poison
name:   Ivysaur
type:   Grass

id:   003
typeTwo:   Poison
name:   Venusaur
type:   Grass

    . . .
```

YAML

YAML Ain't Markup Language (YAML) is a human-friendly data serialization format. It's not as popular as JSON or XML but it was designed to be easily mapped to data types common to most high-level languages. A Python parser implementation called PyYAML is available in PyPI repository and its implementation is very similar to the JSON module.

The first three records of the YAML file (`pokemon.yaml`) look as follows:

```
Pokemon:
  -id     :   001
typeTwo   :   Poison
name      :   Bulbasaur
type      :   Grass
  -id     :   002
typeTwo   :   Poison
```

```
name     :   Ivysaur
type     :   Grass
  -id    :   003
typeTwo  :   Poison
name     :   Venusaur
type     :   Grass
  . . .
```

Getting started with OpenRefine

OpenRefine (formerly known as Google Refine) is a formatting tool very useful in data cleansing, data exploration, and data transformation. It is an open source web application which runs directly in your computer, skipping the problem of uploading your delicate information to an external server.

To start working with OpenRefine just run the application and open a browser in the URL available at `http://127.0.0.1:3333/`.

Refer to *Appendix, Setting Up the Infrastructure*.

Firstly, we need to upload our data and click on **Create Project**. In the following screenshot, we can observe our dataset, in this case, we will use monthly sales of an alcoholic beverages company. The dataset format is an MS Excel (`.xlsx`) worksheet with 160 rows.

We can download the original MS Excel file and the OpenRefine project from the author's GitHub repository available at the following URL:

```
https://github.com/hmcuesta/PDA_Book/tree/master/Chapter2
```

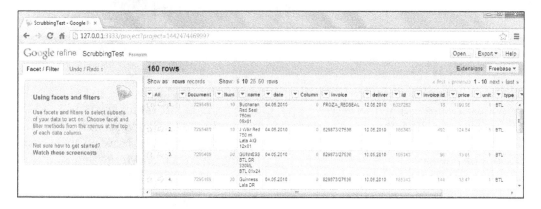

Text facet

Text facet is a very useful tool, similar to filter in a spreadsheet. Text facet groups unique text values into groups. This can help us to merge information and we can see values, which could be spelled in a lot of different ways.

Now we will create a text facet on the **name** column by clicking on that column's drop-down menu and select **Facet | Text Facet**. In the following screenshot we can see the column **name** grouped by its content. This is helpful to see the distribution of elements in the dataset. We will observe the number of choices (43 in this example) and we can sort the information by name or by count.

Clustering

We can cluster all the similar values by clicking on our text facet (refer to the previous screenshot), in this case we find: **Guinness Lata DR 440ml 24x01** and **Guinness Lata DR 440ml 24x01.**, obviously the dot in the second value is a typo. The option **Cluster** allows us to find this kind of dirty data easily. Now we just select the option **Merge?** and define **New Cell Value**, then we click on **Merge Selected & Close** as seen in the following screenshot:

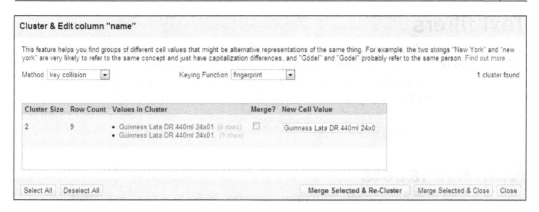

We can play with the parameters of the **Cluster** option, such as changing the **Method** option from **Key collision** to **nearest neighbor**, selecting **Rows in Cluster** or the length variance of choices. Playing with the parameters we can find duplicate items in a data column and more complex misspells, as we can see in the following screenshot, where the values **JW Black Label 750ml 12x01** and **JW Bck Label 750ml 12x01** refer to the same product with a typo in the color.

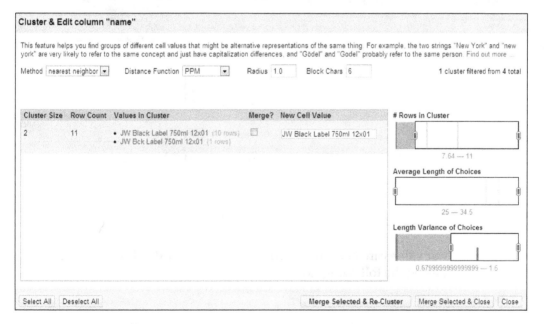

Text filters

We may filter a column by using a specific text string or using a regular expression (Java's regular expressions). We will click on the option **Find** of the column we want to filter and then type our search string in the textbox in the left. For more information about Java's regular expressions visit the following URL:

`http://docs.oracle.com/javase/tutorial/essential/regex/`

Numeric facets

Numeric facet groups numbers into numeric range bins. You can customize numeric facets much the way you can customize text facets. For example, if the numeric values in a column are drawn from a power law distribution (refer to the first row in the following screenshot), then it's better to group them by their logs (refer to the second row in the following screenshot) using the following expression:

```
value.log()
```

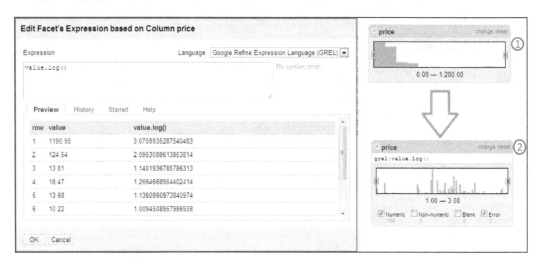

Otherwise, if our values are periodic we could take the **modulus** by the period, to find a pattern, using the following expression:

```
mod(value, 6)
```

We can create a numeric facet from a text by taking the length of the string, using this expression.

```
value.length()
```

Transforming data

In our example, the column **date** uses a special date format 01.04.2013 and we want to replace **.** by **/**. Fixing this is pretty easy using a transform. We need to go to **Column date | Edit Cells | Transform**.

We will write a `replace()` expression as follows:

```
replace(value,".","/")
```

Now just click on the button **OK** to apply the transformation.

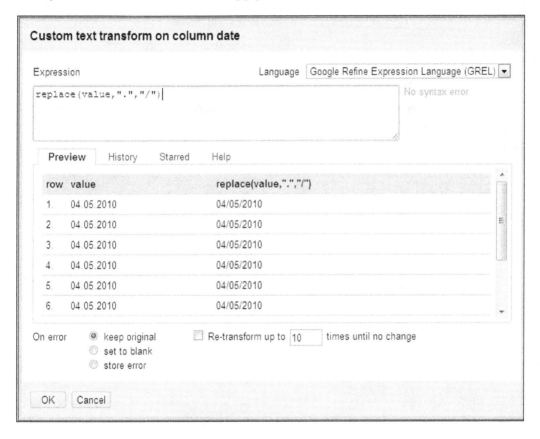

Google Refine Expression Language (GREL) allows us to create complex validations. For example, in simple business logic when the column value reaches 10 units we make a discount of 5 percent, we do this with an `if()` statement and some algebra:

```
if(value>10,value*.95,value)
```

Visit the link for a complete list of functions supported by the GREL available at https://code.google.com/p/ google-refine/wiki/GRELFunctions.

Exporting data

We can export data from an existing OpenRefine project in several formats as follows:

- TSV
- CSV
- Excel
- HTML table

To export the file as a JSON, we need to select the option **Export** and **Templating Export**, where we can specify a JSON template as shown in the following screenshot:

Templating Export

Prefix
```
{
  "rows" : [
```

Row Template
```
{
    "Document" : {{jsonize(cells["Document"].valu
    "Num" : {{jsonize(cells["Num"].value)}},
    "name" : {{jsonize(cells["name"].value)}},
    "date" : {{jsonize(cells["date"].value)}},
    "Column" : {{jsonize(cells["Column"].value)}}
    "invoice" : {{jsonize(cells["invoice"].value)
    "deliver" : {{jsonize(cells["deliver"].value)
    "id" : {{jsonize(cells["id"].value)}},
    "invoice id" : {{jsonize(cells["invoice id"].
    "price" : {{jsonize(cells["price"].value)}},
    "unit" : {{jsonize(cells["unit"].value)}},
    "type" : {{jsonize(cells["type"].value)}},
    "value" : {{jsonize(cells["value"].value)}},
    "value2" : {{jsonize(cells["value2"].value)}}
    "amount" : {{jsonize(cells["amount"].value)}}
```

Row Separator
```
,
```

Suffix
```
  ]
}
```

Right pane:
```
{
  "rows" : [
    {
      "Document" : 7295491,
      "Num" : 10,
      "name" : "Buchanan Red Seal 750ml    06x01",
      "date" : "04/05/2010",
      "Column" : 0,
      "invoice" : "FROZA_REDSEAL",
      "deliver" : "12.05.2010",
      "id" : 6027282,
      "invoice id" : 18,
      "price" : 1190.95,
      "unit" : 1,
      "type" : "BTL",
      "value" : 21437.1,
      "value2" : 32,
      "amount" : 18
    },
    {
      "Document" : 7295489,
      "Num" : 10,
      "name" : "J Wlkr Red 750 ml Lata AIG 12x01",
      "date" : "04/05/2010",
      "Column" : 0,
      "invoice" : "829873/27536",
      "deliver" : "10.05.2010",
      "id" : 185343,
      "invoice id" : 492,
      "price" : 124.54,
      "unit" : 1,
      "type" : "BTL",
```

Reset Template   Export   Cancel

Operation history

We can save all the transformations applied to our dataset just by clicking on the tab **Undo/Redo** and then select **Extract** this will show all the transformations applied to the current dataset (as shown in the following screenshot). Finally, we will copy the generated JSON and we will paste it in a text file.

To apply the transformations to another dataset we just need to open the dataset in OpenRefine, and then go to the tab **Undo/Redo** click on the button **Apply** and copy the JSON from the first project.

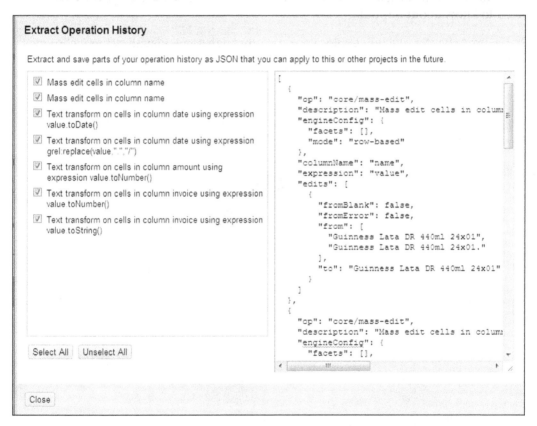

Summary

In this chapter we explored the common datasources and implemented a web scraping example. Next, we introduced the basic concepts of data scrubbing such as statistical methods and text parsing. Then we learned about how to parse the most used text formats with Python. Finally, we presented an introduction to OpenRefine which is an excellent tool for data cleansing and data formatting. Working with data is not just code or clicks, we also need to play with the data and follow our intuition to get our data in great shape. We need to get involved in the knowledge domain of our data to find inconsistencies. Global vision of data helps us to discover what we need to know about our data.

In the next chapter, we will explore our data through some visualization techniques and we will present a fast introduction to D3js.

3
Data Visualization

Sometimes, we don't know how valuable data is until we look at it. In this chapter, we get into a Web Visualization Framework called **D3** (**Data-Driven Documents**) to create visualizations that make complex information easier to understand.

In this chapter we will cover:

- Data-Driven Documents (D3)
- Getting started with D3.js
  - Bar chart
  - Pie chart
  - Scatter plot
  - Line chart
  - Multi-line chart
- Interaction and animation

Exploratory data analysis (**EDA**) as mentioned in *Chapter 1, Getting Started*, is a critical part of the data analysis process because it helps us to detect mistakes, determine relationships and tendencies, or check assumptions. In this chapter, we present some examples of visualization methods for EDA with discrete and continuous data.

The four types of EDA are univariate non-graphical, multivariate non-graphical, univariate graphical, and multivariate graphical. The non-graphical methods refer to the calculation of summary statistics or the outlier detection. In this book, we will focus on the univariate and multivariate graphical models. Using a variety of visualization tools such as bar chars, pie charts, scatter plots, line charts, and multi-line charts, all implemented in D3.js.

In this chapter, we will work with two types of data: discrete data with a list of summarized pokemon types (see *Chapter 2*, *Working with Data*), and hand with continuous data using the historical exchange rates from March 2008 to March 2013. We also explore the creation of a random dataset.

Data-Driven Documents (D3)

D3 is a project featured by the Stanford Visualization Group developed by Mike Bostock.

D3 provides us with web-based visualization, which is an excellent way to deploy information and help us to see things such as proportions, relationships, correlations, and patterns, and discover things previously unknown. Since web browsers provide us with a very flexible and interactive interface in practically any device such as PC, tablet, and smart phone, D3 is an amazing tool for visualization based on data using HTML, JavaScript, SVG, and CSS.

In *Chapter 1*, *Getting Started*, we saw the importance of data visualization and in this chapter, we will present examples in order to understand the use of D3.js. In the following screenshot, we can see the basic structure of an HTML document. D3 is going to be included in a basic script tag or into a JavaScript file (.js):

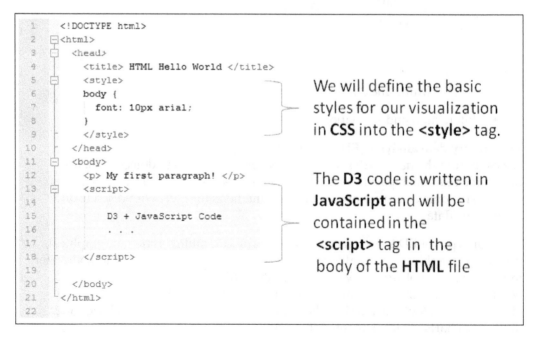

HTML

HyperText Markup Language (HTML) provides the basic skeleton for our visualization. An HTML document will define the structure of our web page, based on a series of tags, which are labels inside angle brackets (`<br/>`) commonly coming in pairs (`<p>...</p>`). D3 will take advantage of the structure of HTML by creating new elements in the document structure, such as adding new `div` tags (which defines a section in a document). We can see the basic structure of an HTML document in the previous screenshot.

 For a complete reference about HTML, please refer to the link `http://www.w3schools.com/html/`.

DOM

Document Object Model (DOM) helps in representing and interacting with objects in HTML documents. Objects in the DOM tree can be addressed and manipulated by programming languages such as Python or JavaScript through the elements (tags) of the web page. D3 will change the structure of the HTML document by accessing the DOM tree either by the element ID or its type.

CSS

Cascading Style Sheets (CSS) can help us to style the web page. A CSS style is based on rules and selectors. We can apply styles to a specific element (tag) through selectors. An example of CSS is shown as follows:

```
<style>
body {
  font: 10px arial;
}
</style>
```

JavaScript

JavaScript is a dynamic scripting programming language typically implemented in the client (web browser). All the code in D3.js is developed with JavaScript. JavaScript will help us to create great visualizations, with full interactivity which can be updated in real time. In D3.js we can link to the library directly (stored in a separate file) with the snippet listed as follows:

```
<script src="http://d3js.org/d3.v3.min.js"></script>
```

SVG

Scalable Vector Graphics (**SVG**) is an XML-based vector image format for two-dimensional graphics. SVG can be directly included in your web page. SVG provides basic shape elements such as rectangle, line, circle, and text to build complicated lines and shapes inside a canvas. Much of the success of D3 is because it implements a wrapper for SVG. With D3 we will not have to modify the XML directly, instead D3 provides an API to help us place our elements (rectangle, circle, line, and so on) in the correct location on the canvas.

Getting started with D3.js

First, download the latest version of D3 from the official website `http://d3js.org/`.

Or, to link directly to the latest release, copy this snippet:

```
<script src="http://d3js.org/d3.v3.min.js"></script>
```

In the basic examples, we can just open our HTML document in a web browser to view it. But when we need to load external data sources, we need to publish the folder on a web server such as Apache, nginx, or IIS. Python provides us with an easy way to run a web server with `http.server`; we just need to open the folder where our D3 files are present and execute the following command in the terminal.

```
$ python3 -m http.server 8000
```

In Windows, you can use the same command by removing the number 3 from python.

```
> python -m http.server 8000
```

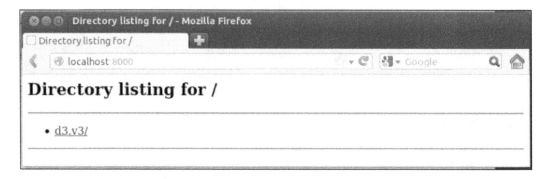

The following examples are based on Mike Bostock's reference gallery, which can be found at `https://github.com/mbostock/d3/wiki/Gallery`.

All the codes and datasets of this chapter can be found in the author's GitHub repository at `https://github.com/hmcuesta/PDA_Book/tree/master/Chapter3`.

Bar chart

Probably, the most common visualization tool is the bar chart. As we can see in the following figure, the horizontal axis (X) represents the category data and the vertical axis (Y) represents a discrete value. We can see the count of pokemon by type with a random sort.

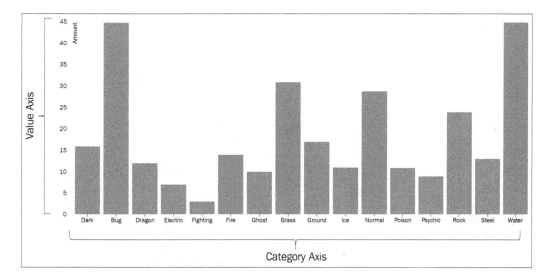

Discrete data are values that can only take certain values; in this case it is the number of pokemon by type.

In the following example, we process the pokemon list in JSON format (see *Chapter 2, Working with Data*) and we get the sum of pokemon by type, sorted by number in ascending order and then save the result in a CSV format. After the data processing, we visualize the result in a bar chart.

The first three records of the JSON file (pokemon.json) look like the following records:

```
[
    {
        "id": " 001",
        "typeTwo": " Poison",
        "name": " Bulbasaur",
        "type": " Grass"
    },
    {
        "id": " 002",
        "typeTwo": " Poison",
        "name": " Ivysaur",
        "type": " Grass"
    },
    {
        "id": " 003",
        "typeTwo": " Poison",
        "name": " Venusaur",
        "type": " Grass"
    },
    . . . ]
```

In this preprocessing stage, we will use Python to turn the JSON file into a CSV format. We will perform an aggregation to get the number of each category of pokemon sorted in ascending order. After we get the resultant CSV file, we will start with the visualization in D3.js. The code for the preprocessing is shown as follows:

```
# We need to import the necessary modules.
import json
import csv
from pprint import pprint
#Now, we define a dictionary to store the result
typePokemon = {}
#Open and load the JSON file.
with open("pokemon.json") as f:
    data = json.loads(f.read())

#Fill the typePokemon dictionary with sum of pokemon by type
    for line in data:
        if line["type"] not in typePokemon:
            typePokemon[line["type"]] = 1
        else:
```

```
              typePokemon[line["type"]]=typePokemon.get(line["type"])+1

#Open in a write mode the sumPokemon.csv file
with open("sumPokemon.csv", "w") as a:
    w = csv.writer(a)

#Sort the dictionary by number of pokemon
#writes the result (type and amount) into the csv file
    for key, value in sorted(typePokemon.items(),
    key=lambda x: x[1]):
        w.writerow([key,str(value)])

 #finally, we use "pretty print" to print the dictionary
    pprint(typePokemon)
```

The result of the preprocessing can be seen in the following table. Each row has two values, the type and the amount of pokemon of a particular type.

Type	Amount
Fighting	3
Electric	7
Psychic	9
Ghost	10
Poison	11
Ice	11
Dragon	12
Steel	13
Fire	14
Dark	16
Ground	17
Rock	24
Normal	29
Grass	31
Water	45
Bug	45

To start working on D3, we need to create a new HTML file with the basic structure (head, style, and body). Next, we will include the styles and the script section as shown in the following steps:

In the CSS, we specified the style for the axis line, the font family, and size for the body and the bar color.

```
<style>
body {
    font: 14px sans-serif;
}
.axis path,
.axis line {
    fill: none;
    stroke: #000;
    shape-rendering: crispEdges;
}
.x.axis path {
    display: none;
}
.bar {
    fill: #0489B1;
}
</style>
```

 We can define the colors in CSS using a hexadecimal code such as #0489B1 instead of the literal name "blue"; in the following link, we can find a color selector http://www.w3schools.com/tags/ref_colorpicker.asp.

Inside the body tag, we need to refer to the library,

```
<body>
<script src="http://d3js.org/d3.v3.min.js"></script>
```

The first thing we need to do is define a new SVG canvas (<svg>) with a width and height of 1000 x 500 pixels, inside the body section of our HTML document.

```
var svg = d3.select("body").append("svg")
    .attr("width", 1000)
    .attr("height", 500)
  .append("g")
    .attr("transform", "translate(50,20)");
```

The transform attribute will help us to translate, rotate, and scale a group of elements (g). In this case we want to translate (move) the position of the margins (left and top) on the canvas with translate("left", "top"). We need to do this because we will need space for the labels in the axis X and Y of our visualization.

Now, we need to open the file sumPokemon.csv and read the values from it. Then, we will create the variable data with two attributes type and amount according to the structure of the CSV file.

The d3.csv method will perform an asynchronous request. When the data is available, a callback function will be invoked. In this case we will iterate the list data and we will convert the amount column to number (d.amount = +d.amount).

```
d3.csv("sumPokemon.csv", function(error, data) {
  data.forEach(function(d) {
    d.amount = +d.amount;
  });
```

Now, we will set a labeled X axis (x.domain) using the map function to get all the type names of the pokemons. Next, we will use the d3.max function to return the maximum value of each type of pokemon for the Y axis (y.domain).

```
x.domain(data.map(function(d) { return d.type; }));
y.domain([0, d3.max(data, function(d) { return d.amount; })]);
```

Now, we will create an SVG Group Element which is used to group SVG elements together with the tag <g>. Then we use the transform function to define a new coordinate system for a set of SVG elements by applying a transformation to each coordinate specified in this set of SVG elements.

```
svg.append("g")
    .attr("class", "x axis")
    .attr("transform", "translate(0,550)")
    .call(xAxis);

svg.append("g")
    .attr("class", "y axis")
    .call(yAxis)
  .append("text")
    .attr("transform", "rotate(-90)")
    .attr("y", 6)
    .attr("dy", ".71em")
    .style("text-anchor", "end")
    .text("Amount");
```

Finally, we need to generate .bar elements and add them to svg, then with the data(data) function, for each value in data, we will call the .enter() function and add a rect element. D3 allows selecting groups of elements for manipulation through the selectAll function.

 In the following link, we can find more information about selections https://github.com/mbostock/d3/wiki/Selections.

```
svg.selectAll(".bar")
    .data(data)
  .enter().append("rect")
    .attr("class", "bar")
    .attr("x", function(d) { return x(d.type); })
    .attr("width", x.rangeBand())
    .attr("y", function(d) { return y(d.amount); })
    .attr("height", function(d) { return height - y(d.amount); });

}); // close the block d3.csv
```

In order to see the result of our visualization, we need to visit the URL http://localhost:8000/bar-char.html, the result is shown in the following screenshot:

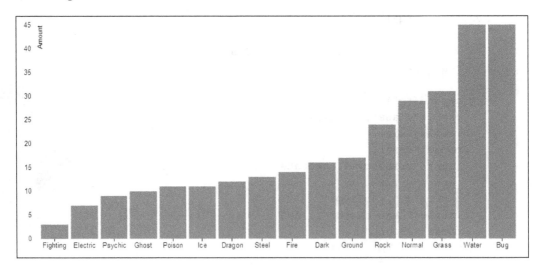

Pie chart

The purpose of the pie chart is to communicate proportions. The sum of all wedges represents a whole (100 percent). Pie charts help us to understand the distribution of proportions in an easy way. In this example, we will use the unordered list of pokemons by type `sumPokemon.csv`, which can be found at `https://github.com/hmcuesta/PDA_Book/tree/master/Chapter3`.

We need to define the font family and size for the labels.

```
<style>
body {
  font: 16px arial;
}
</style>
```

Inside the body tag we need to refer to the library,

```
<body>
<script src="http://d3js.org/d3.v3.min.js"></script>
```

First, we define the size (width, height, and radius) of the work area.

```
var w = 1160,
    h = 700,
    radius = Math.min(w, h) / 2;
```

Now, we will set a range of color that will be used in the chart.

```
var color = d3.scale.ordinal()
    .range(["#04B486", "#F2F2F2", "#F5F6CE", "#00BFFF"]);
```

The function `d3.svg.arc()` creates a circle with an outer radius and an inner radius. See pie charts given later.

```
var arc = d3.svg.arc()
    .outerRadius(radius - 10)
    .innerRadius(0);
```

The function `d3.layout.pie()` specifies how to extract a value from the associated data.

```
var pie = d3.layout.pie()
    .sort(null)
    .value(function(d) { return d.amount; });
```

Now, we select the element body and create a new element `<svg>`.

```
var svg = d3.select("body").append("svg")
    .attr("width", w)
    .attr("height", h)
  .append("g")
    .attr("transform", "translate(" + w / 2 + "," + h / 2 + ")");
```

Next, we need to open the file `sumPokemon.csv` and read the values from the file and create the variable `data` with two attributes `type` and `amount`.

```
d3.csv("sumPokemon.csv", function(error, data) {
  data.forEach(function(d) {
    d.amount = +d.amount;
  });
```

Finally, we need to generate the `.arc` elements and add them to `svg`, then with the `data(pie(data))` function, for each value in data we will call the `.enter()` function and add a `g` element.

```
var g = svg.selectAll(".arc")
    .data(pie(data))
  .enter().append("g")
    .attr("class", "arc");
```

Now, we need to apply the style, color, and labels to the group `g`.

```
g.append("path")
    .attr("d", arc)
    .style("fill", function(d) { return color(d.data.type); });
g.append("text")
    .attr("transform", function(d) { return "translate(" +    arc.
centroid(d) + ")"; })
    .attr("dy", ".60em")
    .style("text-anchor", "middle")
    .text(function(d) { return d.data.type; });
}); // close the block d3.csv
```

In order to see the result of our visualization we need to visit the URL `http://localhost:8000/pie-char.html`. The result is shown in the following screenshot:

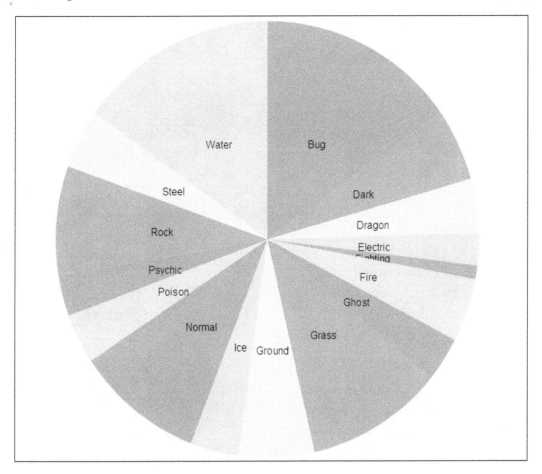

In the following figure, we can see the pie chart with variation in the attribute `inner Radius` of `200` pixels in the function arc.

```
var arc = d3.svg.arc()
    .outerRadius(radius - 10)
    .innerRadius(200);
```

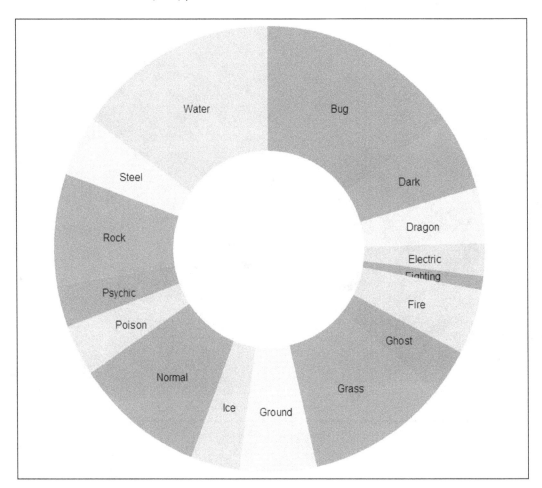

Scatter plot

Scatter plot is a visualization tool based in cartesian space with coordinates of axis X, Y between two different variables, in this case it can be value, categorical, or time represented in data points. Scatter plot allows us to see relationships between the two variables.

In the following figure, we can see a scatter plot, where each data point has two coordinates X and Y. The horizontal axis can take category or time values and in the vertical axis we represent a value.

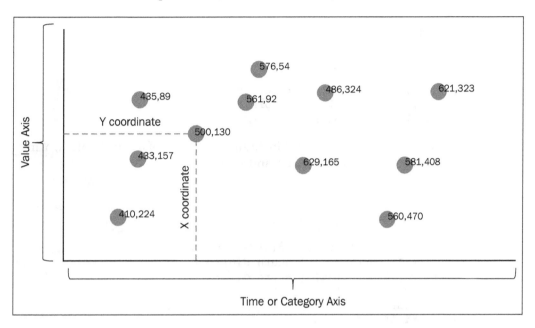

In this example, we generate 20 random points (constrained in a range of 700 x 500) in a bi-dimensional array in JavaScript using the function Math.random() and store the result in the variable data.

```
var data = [];
for(var i=0; i < 20; i++ ){
    var axisX = Math.round(Math.random() * 700);
    var axisY = Math.round(Math.random() * 500);
    data.push([axisX,axisY]);
}
```

Now, we select the element body and create a new element <svg> and define its size.

```
var svg = d3.select("body")
    .append("svg")
    .attr("width", 700)
    .attr("height", 500);
```

We use the selector to create a circle for each data point in the variable `data`, defining the coordinate X as `cx` and the coordinate Y as `cy`, define the radius `r` to `10` pixels and pick a color `fill`.

```
svg.selectAll("circle")
    .data(data)
    .enter()
    .append("circle")
    .attr("cx", function(d) {   return d[0]; })
    .attr("cy", function(d) {   return d[1]; })
    .attr("r", function(d)  {   return 10;   })
    .attr("fill", "#0489B1");
```

Finally, we create the label of each point including the value of a coordinate X, Y in text format. We select a font family, color, and size.

```
svg.selectAll("text")
    .data(data)
    .enter()
    .append("text")
    .text(function(d) {return d[0] + "," + d[1];  })
    .attr("x", function(d) {return d[0];  })
    .attr("y", function(d) {return d[1];  })
    .attr("font-family", "arial")
    .attr("font-size", "11px")
    .attr("fill", "#000000");
```

In the following screenshot, we see the scatter plot that will be seen in our web browser.

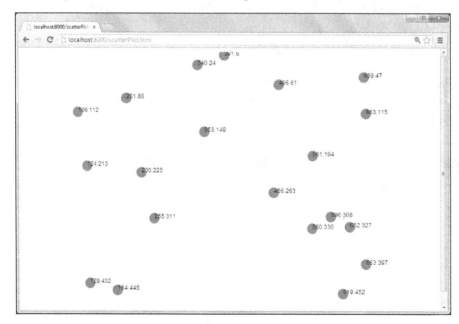

Single line chart

A line chart is a visualization tool that displays continuous data as a series of points connected by a straight line. It is similar to a scatter plot but in this case the points have a logical order and the points are connected — often used for time series' visualization. A time series is a sequence of observations of the physical world in a regular time span. Time series help us to understand trends and correlations. As we can see in the following figure, the vertical axis represents the value data and the horizontal axis represents the time.

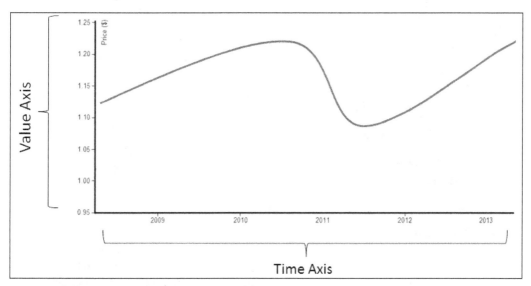

For this example we will use the log of USA/CAD historical exchange rates from March 2008 to March 2013 with 260 records.

In the following link we can find the Historical Exchange Rates log to download `http://www.oanda.com/currency/historical-rates/`.

The first seven records of the CSV file (`line.csv`) look like the following records:

```
date,usd
3/10/2013,1.0284
3/3/2013,1.0254
2/24/2013,1.014
2/17/2013,1.0035
2/10/2013,0.9979
2/3/2013,1.0023
1/27/2013,0.9973
. . .
```

We need to define the font family and size for the labels and the style for the axis line.

```
<style>
body {
  font: 14px sans-serif;
}

.axis path,
.axis line {
  fill: gray;
  stroke: #000;
}
.line {
  fill: none;
  stroke: red;
  stroke-width: 3px;
}
</style>
```

Inside the body tag we need to refer to the library,

```
<body>
<script src="http://d3js.org/d3.v3.js"></script>
```

We will define a format parser for the date value with d3.time.format. In this example, we have the data as: Month/Day/Year — %m/%d/%Y (for example, 1/27/2013). Where, %m represents the month as a decimal number from 01 to 12, %d represents the day of the month as a decimal number from 01 to 31, and %Y represents the year with century as a decimal number.

```
var formatDate = d3.time.format("%m/%d/%Y").parse;
```

 To find out more about time formatting, please refer to the link https://github.com/mbostock/d3/wiki/Time-Formatting/.

Now, we define the X and Y axis with a width of 1000 pixels and height 550 pixels.

```
var x = d3.time.scale()
    .range([0, 1000]);
var y = d3.scale.linear()
    .range([550, 0]);

var xAxis = d3.svg.axis()
    .scale(x)
```

```
      .orient("bottom");

var yAxis = d3.svg.axis()
    .scale(y)
    .orient("left");
```

The `line` element defines a line segment that starts at one point and ends at another.

> In the following link we can find the reference of SVG shapes
> `https://github.com/mbostock/d3/wiki/SVG-Shapes/`.

```
var line = d3.svg.line()
    .x(function(d) { return x(d.date); })
    .y(function(d) { return y(d.usd); });
```

Now, we select the element `body` and create a new element `<svg>` and define its size.

```
var svg = d3.select("body")
    .append("svg")
    .attr("width", 1000)
    .attr("height", 550)
   .append("g")
    .attr("transform", "translate("50,20")");
```

Then, we need to open the file `line.csv` and read the values from the file and create the variable `data` with two attributes `date` and `usd`.

```
d3.csv("line.csv", function(error, data) {
data.forEach(function(d) {
   d.date = formatDate(d.date);
   d.usd = +d.usd;
 });
```

We define the date in horizontal axis (`x.domain`) and in the vertical axis (`y.domain`) and set our value axis with the exchange rates value `usd`.

```
x.domain(d3.extent(data, function(d) { return d.date; }));
y.domain(d3.extent(data, function(d) { return d.usd; }));
```

Finally, we add the groups of points and the labels in the axis.

```
svg.append("g")
    .attr("class", "x axis")
    .attr("transform", "translate(0,550)")
    .call(xAxis);
svg.append("g")
```

```
        .attr("class", "y axis")
        .call(yAxis)
    svg.append("path")
        .datum(data)
        .attr("class", "line")
        .attr("d", line);
}); // close the block d3.csv
```

In the following screenshot we can see the result of the visualization.

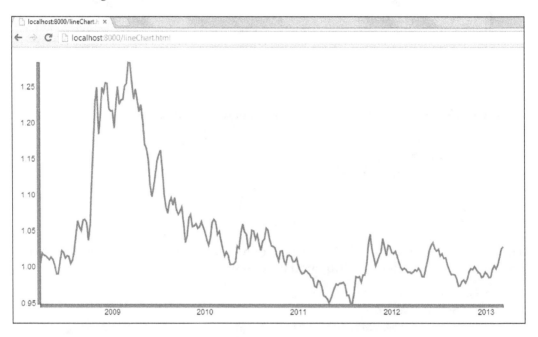

Multi-line chart

In a single variable we can see trends but often we need to compare multiple variables and even find correlations or cluster trends. In this example, we will evolve the last example to work with multi-line chart. In this case we will use data from historical exchange rates from USA, EUR, and GBP.

 In the following link we can find the Historical Exchange Rates log to download http://www.oanda.com/currency/historical-rates/.

The first five records of the CSV file (`multiline.csv`) look like the following records:

```
date,USD/CAD,USD/EUR,USD/GBP
03/10/2013,1.0284,0.7675,0.6651
03/03/2013,1.0254,0.763,0.6609
2/24/2013,1.014,0.7521,0.6512
2/17/2013,1.0035,0.7468,0.6402
02/10/2013,0.9979,0.7402,0.6361
. . .
```

We need to define the font family and size for the labels and the style for the axis line.

```
<style>
body {
    font: 18px sans-serif;
}
.axis path,
.axis line {
    fill: none;
    stroke: #000;
}
.line {
    fill: none;
    stroke-width: 3.5px;
}
</style>
```

Inside the body tag we need to refer to the library,

```
<body>
<script src="http://d3js.org/d3.v3.js"></script>
```

We will define a format parser for the date value with `d3.time.format`.
In this example we have the data as follows Month/Day/Year — %m/%d/%Y
(for example, 1/27/2013).

```
var formatDate = d3.time.format("%m/%d/%Y").parse;
```

Now, we define the X and Y axis with a width of 1000 pixels and height 550 pixels.

```
var x = d3.time.scale()
    .range([0, 1000]);
var y = d3.scale.linear()
    .range([550, 0]);
```

We define an array of color for each line.

```
var color = d3.scale.ordinal()
    .range(["#04B486", "#0033CC", "#CC3300"]);

var xAxis = d3.svg.axis()
    .scale(x)
    .orient("bottom");

var yAxis = d3.svg.axis()
    .scale(y)
    .orient("left");

var line = d3.svg.line()
    .interpolate("basis")
    .x(function(d) { return x(d.date); })
    .y(function(d) { return y(d.currency); });
```

Now, we select the element body and create a new element <svg> and define its size.

```
var svg = d3.select("body")
    .append("svg")
    .attr("width", 1100)
    .attr("height", 550)
  .append("g")
    .attr("transform", "translate("50,20)");
```

Then, we need to open the file multiLine.csv and read the values from the file and create the variable data with two attributes date and color.domain.

```
d3.csv("multiLine.csv", function(error, data) {
  color.domain(d3.keys(data[0]).filter(function(key)
{return key !== "date"; }));
```

Now, we apply the format function to all the column date.

```
data.forEach(function(d) {
  d.date = formatDate(d.date);
});
```

Then, we define currencies as separated array for each color line.

```
var currencies = color.domain().map(function(name) {
  return {
    name: name,
    values: data.map(function(d) {
      return {date: d.date, currency: +d[name]};
```

```
        })
    };
  });

  x.domain(d3.extent(data, function(d) { return d.date; }));
  y.domain
([d3.min(currencies, function(c) { return d3.min(c.values,
function(v) { return v.currency; }); }),
  d3.max(currencies, function(c) { return d3.max(c.values, function(v)
{ return v.currency; }); })
  ]);
```

Now, we add the groups of points as well as the color and labels for each line.

```
svg.append("g")
    .attr("class", "x axis")
    .attr("transform", "translate(0,550)")
    .call(xAxis);
svg.append("g")
    .attr("class", "y axis")
    .call(yAxis)
var country = svg.selectAll(".country")
    .data(currencies)
  .enter().append("g")
  .style("fill", function(d) { return color(d.name); })
  .attr("class", "country");
```

Finally, we add legend to multi-line series chart.

```
country.append("path")
    .attr("class", "line")
    .attr("d", function(d) { return line(d.values); })
    .style("stroke", function(d) { return color(d.name); });
country.append("text").datum(function(d)
{ return {name: d.name, value:  d.values[d.values.length - 1]}; })
    .attr("transform", function(d) {
return "translate("+ x(d.value.date)+","+ y(d.value.currency)+")";
  })
    .attr("x", 10)
  .attr("y", 20)
    .attr("dy", ".50em")
    .text(function(d) { return d.name; });
}); // close the block d3.csv
```

In the following screenshot we can see the result of the visualization.

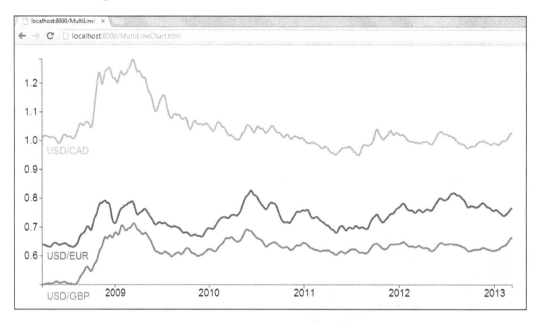

Interaction and animation

D3 provides a good support for interactions, transitions, and animations. In this example, we will focus on the basic way to add transitions and interactions to our visualization. This time we will use a very similar code to that of the bar chart example, in order to demonstrate how easy it is to add interactivity in visualization.

We need to define the font family and size for the labels and the style for the axis line.

```
<style>
body {
  font: 14px arial;
}
.axis path,
.axis line {
  fill: none;
  stroke: #000;
}
.bar {
  fill: gray;
}
</style>
```

Inside the body tag we need to refer to the library,

```
<body>
<script src="http://d3js.org/d3.v3.min.js"></script>
var formato = d3.format("0.0");
```

Now, we define the X and Y axis with a width of 1200 pixels and height 550 pixels.

```
var x = d3.scale.ordinal()
    .rangeRoundBands([0, 1200], .1);

var y = d3.scale.linear()
    .range([550, 0]);

var xAxis = d3.svg.axis()
    .scale(x)
    .orient("bottom");

var yAxis = d3.svg.axis()
    .scale(y)
    .orient("left")
    .tickFormat(formato);
```

Now, we select the element body and create a new element <svg> and define its size.

```
var svg = d3.select("body").append("svg")
    .attr("width", 1200)
    .attr("height", 550)
  .append("g")
    .attr("transform", "translate(20,50)");
```

Then, we need to open the TSV file sumPokemons.tsv and read the values from the file and create the variable data with two attributes type and amount.

```
d3.tsv("sumPokemons.tsv", function(error, data) {
  data.forEach(function(d) {
    d.amount = +d.amount;
  });
```

With the function map we get our categorical values (type of pokemon) for the horizontal axis (x.domain) and in the vertical axis (y.domain) set our value axis with the maximum value by type (in case there is a duplicate value).

```
x.domain(data.map(function(d) { return d.type; }));
y.domain([0, d3.max(data, function(d) { return d.amount; })]);
```

Now, we will create an SVG Group Element which is used to group SVG elements together with the tag <g>. Then we use the `transform` function to define a new coordinate system for a set of SVG elements by applying a transformation to each coordinate specified in this set of SVG elements.

```
svg.append("g")
    .attr("class", "x axis")
    .attr("transform", "translate(0," + height + ")")
    .call(xAxis);

svg.append("g")
    .attr("class", "y axis")
    .call(yAxis)
```

Now, we need to generate `.bar` elements and add them to `svg`; then with the `data(data)` function, for each value in data we will call the `.enter()` function and add a `rect` element. D3 allows selecting groups of elements for manipulation through the `selectAll` function.

We want to highlight any bar just by clicking on it. First, we need to define a click event with `.on('click',function)`. Next we will define the change of style for the bar highlighted with `.style('fill','red');`. In the following figure we can see the highlighted bars Bug, Fire, Ghost, and Grass. Finally, we are going to set a simple animation using a transition `transition().delay` with a delay between the appearing of each bar. (See the next bar chart).

For a complete reference about Selections follow the link
https://github.com/mbostock/d3/wiki/Selections/.

```
svg.selectAll(".bar")
    .data(data)
    .enter().append("rect")
    .on('click', function(d,i) {
    d3.select(this).style('fill','red');
})
    .attr("class", "bar")
    .attr("x", function(d) { return x(d.type); })
```

For a complete reference about Transitions follow the link
https://github.com/mbostock/d3/wiki/Transitions/.

```
    .attr("width", x.rangeBand())
    .transition().delay(function (d,i){ return i * 300;})
```

```
    .duration(300)
    .attr("y", function(d) { return y(d.amount); })
    .attr("height", function(d) { return 550 - y(d.amount);})
    ;
}); // close the block d3.tsv
```

 All the codes and datasets of this chapter can be found in the author GitHub repository at the link `https://github.com/hmcuesta/PDA_Book/tree/master/Chapter3`.

In the following figure we can see the result of the visualization.

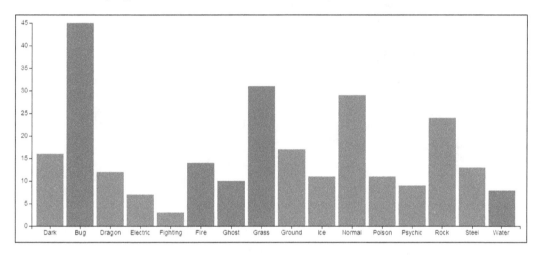

Summary

The visualization is an efficient way to find frequent patterns or relationships in a dataset. In this chapter, we were introduced to a number of basic graphs implemented with `D3.js`. We discussed the most popular visualization techniques for discrete and continuous data. We explored the relation between variables and how variables work over time.

Finally, we presented how to integrate basic user interaction and simple animation, which will be used widely in the following chapters.

In the next chapter, we will introduce a variety of data analysis projects using machine learning algorithms as well as visualization tools.

4
Text Classification

This chapter provides a brief introduction to text classification and also provides you with an example of the **Naïve Bayes** algorithm, developed from scratch, in order to explain how to turn an equation into a code.

In this chapter we will cover:

- Learning and classification
- Bayesian classification
- Naïve Bayes algorithm
- E-mail subject line tester
- The data
- The algorithm
- Classifier accuracy

Learning and classification

When we want to automatically identify to which category a specific value (categorical value) belongs, we need to implement an algorithm that can predict the most likely category for the value, based on the previous data. This is called Classification. In the words of Tom Mitchell:

> *"How can we build computer systems that automatically improve with experience, and what are the fundamental laws that govern all learning processes?"*

The keyword here is learning (supervised learning in this case), and also how to train an algorithm to identify categorical elements. The common examples are **spam classification, speech recognition, search engines, computer vision,** and **language detection**; but there are a large number of applications for a classifier. We can find two kinds of problems in classification. The **binary classification** is where we have only two categories (spam or not spam) and **multiclass classification** is where many categories are involved (for example, opinions can be positive, neutral, negative, and so on). We can find several algorithms for classification, the most frequently used are **support vector machines, neural networks, decision trees, Naïve Bayes,** and **hidden Markov models**. In this chapter, we will implement a probabilistic classification using Naïve Bayes algorithm, but in the following chapters we will implement several other classification algorithms for a variety of problems.

The general steps involved in supervised classification are shown in the following figure. First we will collect training data (previously classified), then we will perform feature extraction (relevant features for the categorization). Next, we will train the algorithm with the features vector. Once we get our trained classifier, we may insert new strings, extract their features, and send them to the classifier. Finally, the classifier will give us the most likely class (category) for the new string.

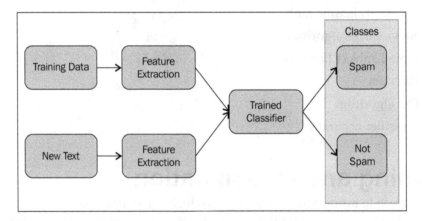

Additionally we will test the classifier accuracy by using a hand-classified test set. Due to this, we will split the data into two sets, the training data and the test data.

Bayesian classification

Probabilistic classification is a practical way to draw inferences based on data, using statistical inference to find the best class for a given value. Given the probability distribution, we can select the best option with the highest probability. The Bayes theorem is the basic rule to draw inferences. The Bayes theorem allows us to update the likelihood of an event, given the new data or observations. In other words, it allows us to update the prior probability $P(A)$ to the posterior probability $P(A \mid B)$. The prior probability is given by the likelihood before the data is evaluated and the posterior probability is assigned after the data is taken into account. The following expression represents the Bayes theorem:

$$P(A \mid B) = \frac{P(B \mid A)P(A)}{P(B)}$$

$P(A \mid B)$ = The conditional probability of A given B

Naïve Bayes algorithm

Naïve Bayes is the simplest classification algorithm among Bayesian classification methods. In this algorithm, we simply need to learn the probabilities by making the assumption that the attributes A and B are independent, that's why this model is defined as an independent feature model. Naïve Bayes is widely used in text classification because the algorithm can be trained easily and efficiently. In Naïve Bayes we can calculate the probability of a condition A given B (described as $P(A \mid B)$), if we already know the probability of B given A (described as $P(B \mid A)$), and additionally the probability of A (described as $(P(A))$) and the probability of B (described as $P(B)$) individually, as is shown in the preceding Bayes Theorem.

E-mail subject line tester

An e-mail subject line tester is a simple program, which will define if a certain subject line in an e-mail is spam or not. In this chapter, we will program a Naïve Bayes classifier from scratch. The example will classify if a subject line is spam or not using a very simple code. This will be done by breaking the subject lines into a list of relevant words, which will be used as the features vectors in the algorithm. In order to do this, we will use the **SpamAssassin** public dataset. SpamAssasin includes three categories; spam, easy ham, and hard ham. In this case, we will create a binary classifier with two classes spam and not spam (easy ham).

There are several features that we can use for our classifier such as the precedence, the language, and the use of upper case. We will keep things simple and use the frequency of only those words which consist of more than three characters, avoiding words such as **The** or **RT**, when training the algorithm.

We will implement the Bayes rule, using the words and categories, as shown in the following equation:

$$P(word \mid category) = \frac{P(category \mid word)\,P(word)}{P(category)}$$

For more information about probability distributions, please refer to http://en.wikipedia.org/wiki/Probability_distribution.

Here, we have two classes in the categories which represents if a subject line is spam or not. We need to split the texts into a list of words in order to get the likelihood of each word. Once we know the probability of each word, we need to multiply the probabilities for each category as shown in the following equation:

$$P(category \mid word_1, word_2, ..., word_n) = P(category) x \pi_i P(word \mid category)$$

In other words, we multiply the likelihood of each word *P(word | category)* of the subject line and the probability of the category *P(category)*.

For training the algorithm, we need to provide with some prior examples. In this case, we will use the `training()` function that needs a dictionary of subject line and category, as we can see in the following table:

Subject line	Category
Re: Tiny DNS Swap	nospam
Save up to 70% on international calls!	nospam
[Ximian Updates] Hyperlink handling in Gaim allows arbitrary code to be executed	nospam
Promises.	nospam
Life Insurance - Why Pay More?	spam
[ILUG] Guaranteed to lose 10-12 lbs in 30 days 10.206	spam

The data

We can find the spam dataset at `http://spamassassin.apache.org/`.

In the following screenshot, we can see the easy ham (not spam) folder with 2551 files:

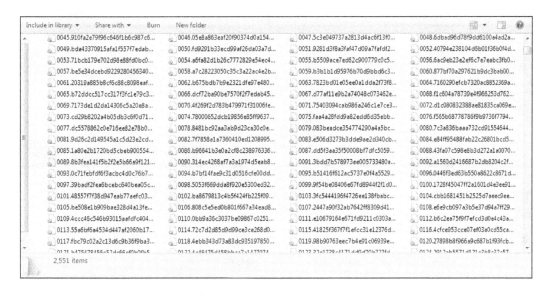

The spam text looks very much similar to the following screenshot, and may include HTML tags and plain text. In this case, we are only interested in the subject line so we need to write a code to obtain the subject from all the files.

```
 1    From smilee1313@eudoramail.com  Mon Aug 26 18:32:20 2002
 2    Return-Path: <smilee1313@eudoramail.com>
 3    Delivered-To: zzzz@localhost.spamassassin.taint.org
 4    Received: from localhost (localhost [127.0.0.1])
 5        by phobos.labs.spamassassin.taint.org (Postfix) with ESMTP id 4ABDD43F9B
 6        for <zzzz@localhost>; Mon, 26 Aug 2002 13:32:20 -0400 (EDT)
 7    Received: from mail.webnote.net [193.120.211.219]
 8        by localhost with POP3 (fetchmail-5.9.0)
 9        for zzzz@localhost (single-drop); Mon, 26 Aug 2002 18:32:20 +0100 (IST)
10    Received: from proxy-server.argogroupage.gr (mail.argogroupage.gr [195.97.102.134])
11        by webnote.net (8.9.3/8.9.3) with ESMTP id SAA27069
12        for <zzzz@spamassassin.taint.org>; Mon, 26 Aug 2002 18:30:18 +0100
13    Message-Id: <200208261730.SAA27069@webnote.net>
14    Received: from smtp0291.mail.yahoo.com (210.83.114.125 [210.83.114.125]) by proxy-s
15        id QP7CPKKZ; Sat, 24 Aug 2002 02:20:16 +0300
16    Date: Sat, 24 Aug 2002 07:08:34 +0800
17    From: "Jeannie Quiroz" <smilee1313@eudoramail.com>
18    X-Priority: 3
19    To: zzzz@netcomuk.co.uk
20    Cc: zzzz@spamassassin.taint.org, yyyy@netvision.net.il, yyyy@nevlle.net,
21        zzzz@news4.inlink.com
22    Subject: zzzz,Increase your breast size. 100% safe!
23    Mime-Version: 1.0
24    Content-Type: text/plain; charset=us-ascii
25    Content-Transfer-Encoding: 7bit
26
27    ===================================
28
29    Guaranteed to increase, lift and firm your
30    breasts in 60 days or your money back!!
31
32    100% herbal and natural.  Proven formula since
33    1996.  Increase  your bust by 1 to 3 sizes within 30-60
34    days and be all natural.
```

Hyper Text Markup Language file   length : 2118   lines : 57

This example will show how to preprocess the SpamAssassin data, using **Python**, in order to collect all the subject lines from the e-mails.

First, we need to import the os module, in order to get the list of filenames using the listdir function from the \spam and \easy_ham folders:

```
import os
files = os.listdir(r" \spam")
```

We will need a new file to store the subject lines and category (spam or not spam), but this time we will use a comma as a separator:

```
with open("SubjectsSpam.out","a") as out:
    category = "spam"
```

Now, we will parse each file and get the subject. Finally, we write the subject and the category in a new file, and delete all the commas from the subject lines (`line.replace(",", "")`) to skip future troubles with the CSV format:

```
for fname in files:
    with open("\\spam\\" + fname) as f:
        data = f.readlines()
        for line in   data:
            if line.startswith("Subject:"):
                line.replace(",", "")
                print(line)
                out.write("{0}, {1} \n".format(line[8:-1], category))
```

We use `line[8:-1]` to skip the word `Subject:` (8-characters long) and the enter at the end of the line (`-1`):

`Output:`

`>>>Hosting from ?6.50 per month`

`>>>Want to go on a date?`

`>>>[ILUG] ilug,Bigger, Fuller Breasts Naturally In Just Weeks`

`>>> zzzz Increase your breast size. 100% safe!`

We will keep the spam and not spam in different files, to play with the size of the training sets and test sets. Usually, more data in the training set means better performance of the algorithm but in this case we will try to find an optimal threshold of the training set size.

 All the codes and datasets of this chapter can be found in the author's GitHub repository at `https://github.com/hmcuesta/PDA_Book/tree/master/Chapter4`.

The algorithm

We use the `list_words()` function to get a list of unique words which are more than three-characters long and in lower case:

```
def list_words(text):
  words = []
  words_tmp = text.lower().split()
  for w in words_tmp:
    if w not in words and len(w) > 3:
      words.append(w)
  return words
```

 For a more advanced term-document matrix, we can use Python's textmining package from `https://pypi.python.org/pypi/textmining/1.0`.

The `training()` function creates variables to store the data needed for the classification. The `c_words` variable is a dictionary with the unique words and its number of occurrences in the text (frequency) by category. The `c_categories` variable stores a dictionary of each category and its number of texts. Finally, `c_text` and `c_total_words` store the total count of texts and words respectively:

```
def training(texts):
  c_words ={}
  c_categories ={}
  c_texts = 0
  c_total_words =0
  #add the classes to the categories
  for t in texts:
    c_texts = c_texts + 1
    if t[1] not in c_categories:
      c_categories[t[1]] = 1
    else:
      c_categories[t[1]]= c_categories[t[1]] + 1

  #add the words with list_words() function
  for t in texts:
    words = list_words(t[0])

    for p in words:
      if p not in c_words:
        c_total_words = c_total_words +1
        c_words[p] = {}
```

```
       for c in c_categories:
           c_words[p][c] = 0
      c_words[p][t[1]] = c_words[p][t[1]] + 1


    return (c_words, c_categories, c_texts, c_total_words)
```

The `classifier()` function applies the Bayes rule and classifies the subject into one of the two categories, that is, either spam or not spam. The function also needs the four variables from the `training()` function:

```
def classifier(subject_line, c_words, c_categories, c_texts, c_tot_
words):
  category =""
  category_prob = 0

  for c in c_categories:
    #category probability
    prob_c = float(c_categories[c])/float(c_texts)
    words = list_words(subject_line)
    prob_total_c = prob_c
    for p in words:
      #word probability
      if p in c_words:
        prob_p= float(c_words[p][c])/float(c_tot_words)
        #probability P(category|word)
        prob_cond = prob_p/prob_c
        #probability P(word|category)
        prob =(prob_cond * prob_p)/ prob_c
        prob_total_c = prob_total_c * prob

    if category_prob < prob_total_c:
      category = c
      category_prob = prob_total_c
  return (category, category_prob)
```

Finally, we will read the `training.csv` file, which contains the training dataset, in this case, 100 spam and 100 not spam subject lines:

```
if __name__ == "__main__":
  with open('training.csv') as f:
    subjects = dict(csv.reader(f, delimiter=','))
  words,categories,texts,total_words = training(subjects)
```

Now, to check if everything is working correctly, we test the classifier with one subject line:

```
clase = classifier("Low Cost Easy to Use Conferencing"
  , words,categories,texts,total_words)

print("Result: {0} ".format(clase))
```

We can see the result in the python console, and from the result we can see the classifier is working correctly so far:

```
>>> Result: ('spam', 0.18518518518518517)
```

We can see the complete code of the Naïve Bayes classifier listed as follows:

```
import csv
def list_words(text):
  words = []
  words_tmp = text.lower().split()
  for p in words_tmp:
    if p not in words and len(p) > 3:
      words.append(p)
  return words

def training(texts):
  c_words ={}
  c_categories ={}
  c_texts = 0
  c_tot_words =0

  for t in texts:
    c_texts = c_texts + 1
    if t[1] not in c_categories:
      c_categories[t[1]] = 1
    else:
      c_categories[t[1]]= c_categories[t[1]] + 1

  for t in texts:
    words = list_words(t[0])

  for p in words:
    if p not in c_words:
      c_tot_words = c_tot_words +1
      c_words[p] = {}
      for c in c_categories:
        c_words[p][c] = 0
```

```
      c_words[p][t[1]] = c_words[p][t[1]] + 1

  return (c_words, c_categories, c_texts, c_tot_words)


def classifier(subject_line, c_words, c_categories, c_texts,
                                            c_tot_words):
  category =""
  category_prob = 0

  for c in c_categories:

    prob_c = float(c_categories[c])/float(c_texts)
    words = list_words(subject_line)
    prob_total_c = prob_c
    for p in c_words:

      if p in words:
        prob_p= float(c_words[p][c])/float(c_tot_words)
        prob_cond = prob_p/prob_c
        prob =(prob_cond * prob_p)/ prob_c
        prob_total_c = prob_total_c * prob

      if category_prob < prob_total_c:
        category = c
        category_prob = prob_total_c
    return (category, category_prob)


if __name__ == "__main__":

  with open('training.csv') as f:
    subjects = dict(csv.reader(f, delimiter=','))

  w,c,t,tw = training(subjects)
  clase = classifier("Low Cost Easy to Use Conferencing"
    ,w,c,t,tw)
  print("Result: {0} ".format(clase))

  with open("test.csv") as f:
    correct = 0
    tests = csv.reader(f)
    for subject in test:
      clase = classifier(subject[0],w,c,t,tw)
      if clase[1] =subject[1]:
        correct += 1
    print("Efficiency : {0} of 100".format(correct))
```

Classifier accuracy

Now we need to test our classifier with a bigger test set. In this case, we will randomly select 100 subjects; 50 spam and 50 not spam. Finally, we will count how many times the classifier chose the correct category:

```
with open("test.csv") as f:
  correct = 0
  tests = csv.reader(f)
  for subject in test:
    clase = classifier(subject[0],w,c,t,tw)
    if clase[1] =subject[1]:
      correct += 1
  print("Efficiency : {0} of 100".format(correct))
```

In this case, the efficiency is 82 percent:

```
>>> Efficiency: 82 of 100
```

 We can find out of the box implementations of Naïve Bayes classifier such as the `NaiveBayesClassifier` function in the NLTK package for Python. NLTK provides a very powerful natural language toolkit and we can download it from `http://nltk.org/`.

In *Chapter 11, Sentiment Analysis of Twitter Data*, we present a more sophisticated version of Naïve Bayes classifier to perform a sentiment analysis.

In this case, we will find an optimal-size threshold for the training set. We try a different number of random subject lines. In the following figure, we can see four of the seven tests and the classification rate (accuracy) of the algorithm. In all cases, we use the same test set of 100 elements and we are using the same number of e-mail subject lines of each category.

The results of four tests are as follows:

- Test 1: 82 percent with a training set of 200 elements
- Test 2: 85 percent with a training set of 300 elements
- Test 5: 87 percent with a training set of 500 elements
- Test 7: 92 percent with a training set of 800 elements

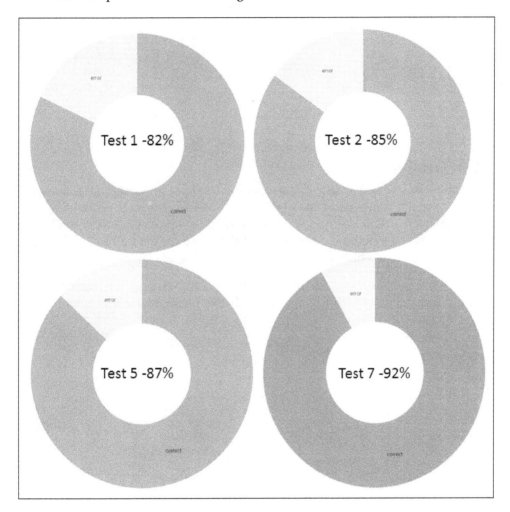

As we can see in the following figure, for this specific example, the maximum accuracy is 92 percent and the optimal number of texts in the training set is 700. After 700 texts in the training set, the accuracy of the classifier doesn't see significant improvement.

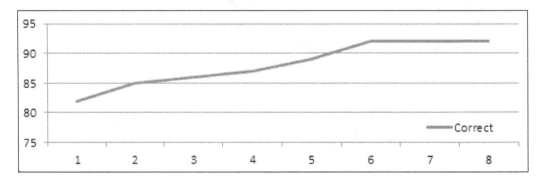

Summary

In this chapter, we created a basic but useful e-mail subject line tester. This chapter provided a guide to code a basic Naïve Bayes classifier from scratch without any external library, in order to demonstrate how easy it is to program a machine-learning algorithm. We also defined the maximum size threshold for the training set and got an accuracy of 92 percent, which, for this basic example, is quite good.

In the following chapters, we will introduce more complex machine learning algorithms, using the `mlpy` library, and we will also present how to extract more sophisticated features.

5
Similarity-based Image Retrieval

A big part of the data that we work with is presented as an image, drawing, or photo. In this chapter, we will implement a **similarity-based image retrieval** without the use of any metadata or concept-based image indexing. We will work with distance metric and dynamic warping to retrieve the most similar images.

In this chapter we will cover:

- Image similarity search
- Dynamic time warping
- Processing the image dataset
- Implementing DTW
- Analyzing the results

Image similarity search

While comparing two or more images, the first question that comes to our mind is what makes an image similar to another? We can say that one image is equal to another if all their pixels match. However, a small change in the light, angle, or rotation of the camera represents a big change in the numerical values of the pixels. Finding ways to define if two images are similar is the main concern of services such as Google Search by Image or TinEye, where the user uploads an image instead of providing keywords or descriptions as search criteria.

Humans have natural mechanisms to detect patterns and similarity. Comparing images at content or semantic level is a difficult problem and an active research field in computer vision, image processing, and pattern recognition. We can represent an image as a matrix (two-dimensional array), in which each position of the matrix represents the intensity or the color of the image. However, any change in the lighting, camera angle, or rotation means a large numerical shift in the matrix. The question that comes to our mind is how can we measure similarity between matrices? To address these problems the data analysis implements several **content-based image retrieval (CBIR)** tools such as comparison of wavelets, Fourier analysis, or pattern recognition with neural networks. However, these methods cause a loss of a lot of the image information, or need an extensive training similar to the neural networks. The most used method is the **description-based image retrieval** using metadata associated with the images, but in an unknown dataset, this method is not effective.

In this chapter, we used a different approach, taking advantage of the **elastic matching** of a time series, which is a method widely used in voice recognition and time series comparison. For the purposes of this chapter, we understand the time series as a sequence of pixels. The trick is to turn the pixels of the image into a numerical sequence, as is shown in the following figure:

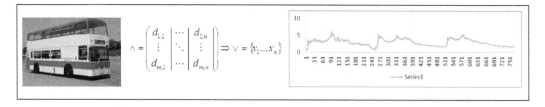

Dynamic time warping (DTW)

Dynamic time warping (DTW) is an elastic matching algorithm used in pattern recognition. DTW finds the optimal warp path between two time series. DTW is used as a distance metric, often implemented in speech recognition, data mining, robotics, and in this case image similarity.

The distance metric measures how far are two points A and B from each other in a geometric space. We commonly use the Euclidian distance which draws a direct line between the pair of points. In the following figure, we can see different kinds of paths between the points A and B such as the **Euclidian distance** (with the arrow) but also we see the **Manhattan (or taxicab) distance** (with the dotted lines), which simulate the way a New York taxi moves through the buildings.

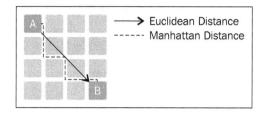

DTW is used to define similarity between time series for classification, in this example, we will implement the same metric with sequences of pixels. We can say that if the distance between the sequence A and B is small, these images are similar. We will use Manhattan distance between the two series to sum of the squared distances. However, we can use other distances metrics such as **Minkowski** or **Euclidean**, depending on the problem at hand.

In the following figure we can observe warping between two time series:

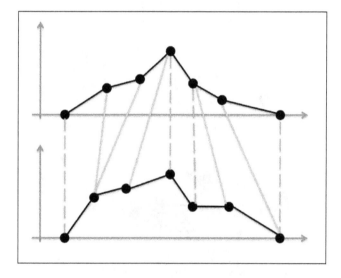

 Distance metrics are formulated in the Taxicab geometry proposed by Hermann Minkowski, for more information about it visit http://taxicabgeometry.net/.

The example of this chapter will use `mlpy`, which is a Python module for Machine Learning, built on top of `numPy` and `sciPy`. The `mlpy` library implements a version of DTW that can be found at `http://mlpy.sourceforge.net/docs/3.4/dtw.html`.

See *Appendix, Setting Up the Infrastructure*, for complete instructions on how to install `mlpy` library.

In the paper *Direct Image Matching by Dynamic Warping*, Hansheng Lei and Venu Govindaraju implement a DTW for image matching, finding an optimal pixel-to-pixel alignment, and prove that DTW is very successful in the task.

In the following figure we can observe a cost matrix with the minimum distance warp path traced through it to indicate the optimal alignment:

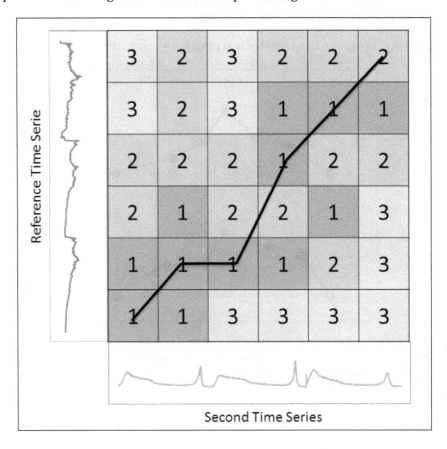

Processing the image dataset

The image set used in this chapter is the **Caltech-256**, obtained from the Computational Vision Lab at CALTECH. We can download the collection of all 30607 images and 256 categories from `http://www.vision.caltech.edu/Image_Datasets/Caltech256/`.

In order to implement the DTW, first we need to extract a time series (pixel sequences) from each image. The time series will have a length of 768 values adding the 256 values of each color in the **RGB (Red, Green, and Blue)** color model of each image. The following code implements the `Image.open("Image.jpg")` function and cast into an array, then simply add the three vectors of color in the list:

```
from PIL import Image
img = Image.open("Image.jpg")
arr = array(img)
list = []
for n in arr: list.append(n[0][0]) #R
for n in arr: list.append(n[0][1]) #G
for n in arr: list.append(n[0][2]) #B
```

Pillow is a PIL fork by Alex Clark, compatible with Python 2.x and 3.x. PIL is the Python Imaging Library by Fredrik Lundh. In this chapter, we will use Pillow due to its compatibility with Python 3.2 and can be downloaded from `https://github.com/python-imaging/Pillow`.

Implementing DTW

In this example, we will look for similarity in 684 images from 8 categories. We will use four imports of `PIL`, `numpy`, `mlpy`, and `collections`:

```
from PIL import Image
from numpy import array
import mlpy
from collections import OrderedDict
```

First, we need to obtain the time series representation of the images and store it in a dictionary (data) with the number of the image and its time series as `data[fn] = list`.

 The performance of this process will lie in the number of images processed, so beware of the use of memory with large datasets.

```
data = {}

for fn in range(1,685):
    img = Image.open("ImgFolder\\{0}.jpg".format(fn))
    arr = array(img)
    list = []
    for n in arr: list.append(n[0][0])
    for n in arr: list.append(n[0][1])
    for n in arr: list.append(n[0][2])
    data[fn] = list
```

Then, we need to select an image for the reference, which will be compared with all the other images in the `data` dictionary:

```
reference = data[31]
```

Now, we need to apply the `mlpy.dtw_std` function to all the elements and store the distance in the `result` dictionary:

```
result ={}
for x, y in data.items():
    #print("{0} --------------- {1}".format(x,y))
    dist = mlpy.dtw_std(reference, y, dist_only=True)
    result[x] = dist
```

Finally, we need to sort the result in order to find the closest elements with the `OrderedDict` function and print the ordered result:

```
sortedRes = OrderedDict(sorted(result.items(), key=lambda x:
    x[1]))
for a,b in sortedRes.items():
    print("{0}-{1}".format(a,b))
```

In the following screenshot, we can see the result and we can observe that the result is accurate with the first element (reference time series). The first result presents a distance of **0.0** because it's exactly similar to the image we used as a reference.

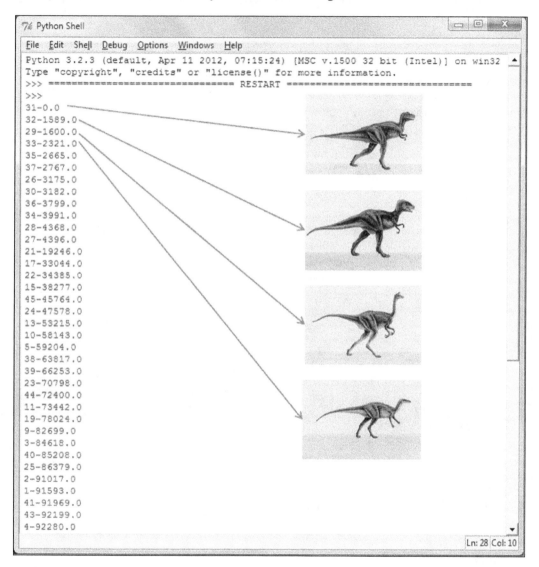

All the codes and datasets of this chapter may be found in the author's GitHub repository at `https://github.com/hmcuesta/PDA_Book/tree/master/Chapter5`.

We can see the complete code as follows:

```
from PIL import Image
from numpy import array
import mlpy
from collections import OrderedDict

data = {}

for fn in range(1,685):
    img = Image.open("ImgFolder\\{0}.jpg".format(fn))
    arr = array(img)
    list = []
    for n in arr: list.append(n[0][0])
    for n in arr: list.append(n[0][1])
    for n in arr: list.append(n[0][2])
    data[fn] = list
reference = data[31]

result ={}

for x, y in data.items():
    #print("{0} --------------- {1}".format(x,y))
    dist = mlpy.dtw_std(reference, y, dist_only=True)
    result[x] = dist

sortedRes = OrderedDict(sorted(result.items(), key=lambda x:
    x[1]))
for a,b in sortedRes.items():
    print("{0}-{1}".format(a,b))
```

Analyzing the results

This example presents a basic implementation that can be adapted in several cases such as 3D object recognition, face recognition, or image clustering. The goal of this chapter is to present how we can easily compare time series without any previous training, in order to find the similarity between images. In this section we present seven cases and will analyze the results.

In the following figure, we can see the first three searches and can observe a good accuracy in the result, even in case of the bus the result displays the result elements in different angles, rotation, and colors:

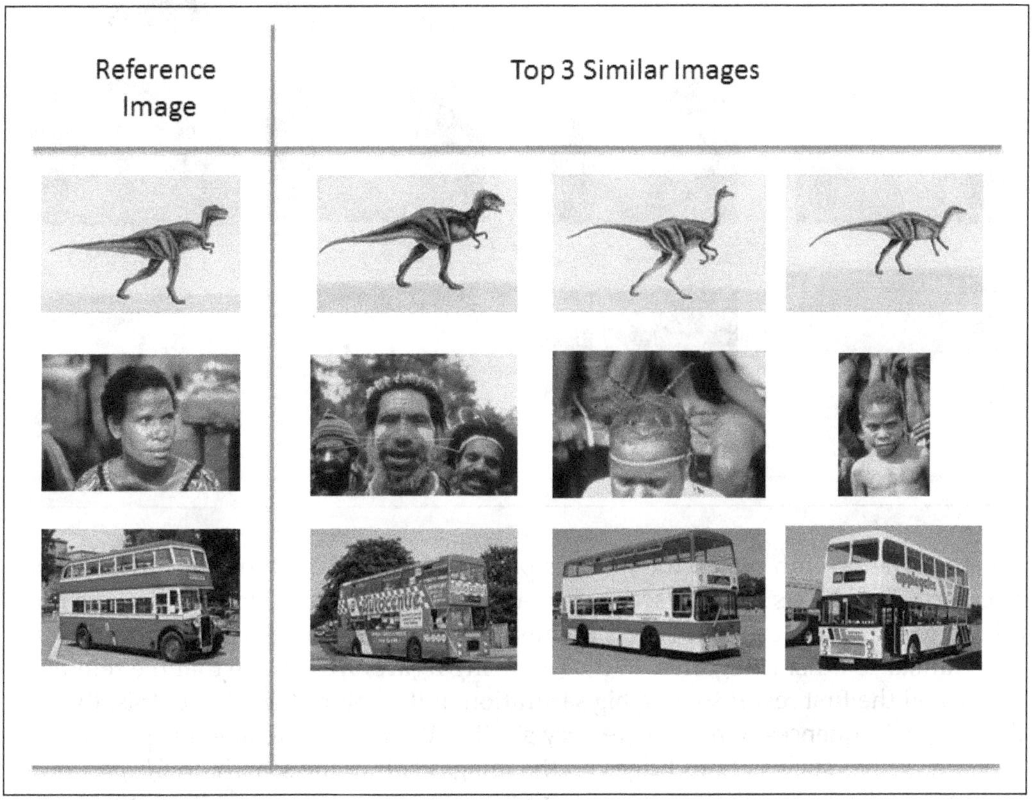

In the following figure, we see the fourth, fifth, and sixth search, and we can observe that the algorithm performs well with an image that has a good contrast in colors.

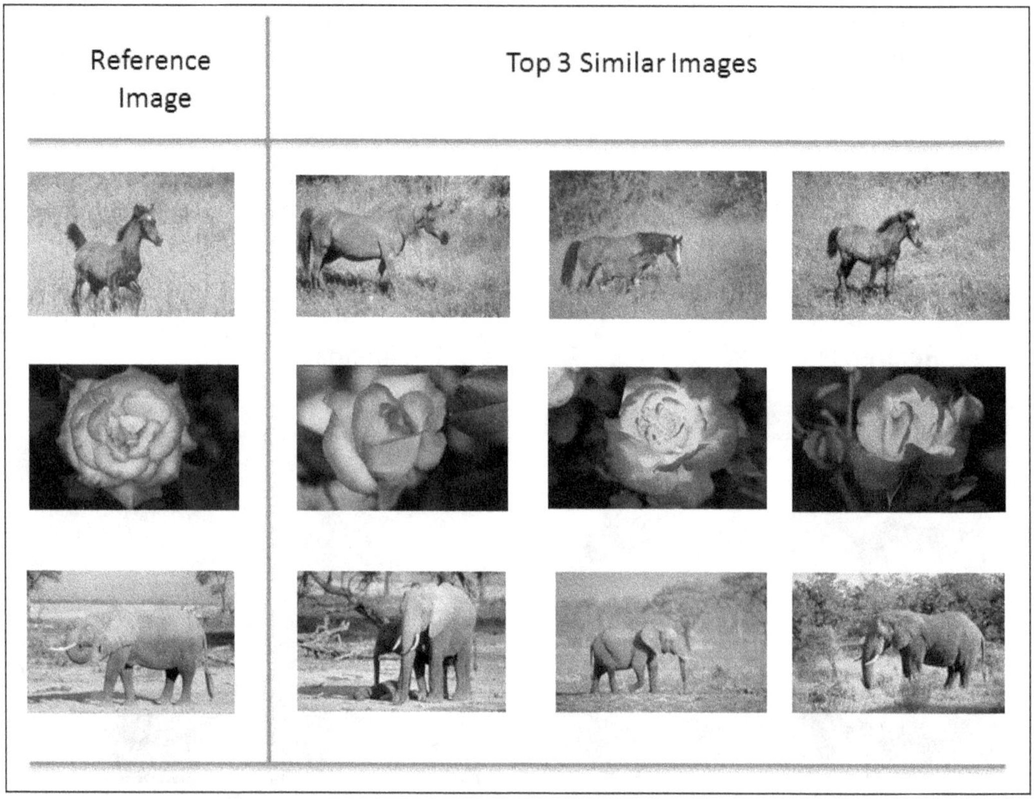

Reference Image	Top 3 Similar Images		

In case of the seventh search the result is poor, and in similar cases when the references time series is a landscape or a building, the result are images that are not related to the search criteria. This is because the RGB color model of the time series is very similar to other categories. In the following figure, we can see that the reference image and the first result share a big saturation of the color blue. Due to this, their time series (sequences of pixels) are very similar. We may overcome this problem by using a filter such as **Find Edges** on the images before the search. In *Chapter 14, Online Data Analysis with IPython and Wakari*, we present the use of filters, operations, and transformations for image processing using PIL.

| Reference Image | Top 3 Similar Images | | |

In the following table we can see the result of the complete set of tests:

Categories	Number of Images	% First Result Right	% Second Result Right
Dinosaurs	102	99	99
African people	85	98	95
Bus	56	98	90
Horse	122	92	88
Roses	95	96	92
Elephants	36	98	87
Landscape	116	60	52
Buildings	72	50	45

Summary

In this chapter, we introduced the dynamic time warping (DTW) algorithm, which is an excellent tool to find similarity between time series without any previous training. We presented an implementation of DTW to find similarity between a set of images, which worked very well in most cases. This method can be used for several other problems in a variety of areas such as robotics, computer vision, speech recognition, and time series analysis. We also saw how to turn an image into a time series with the PIL library. Finally we learned how to implement DTW with the mlpy library. In the next chapter, we will present how simulation can help us in the data analysis and how to model pseudo-random events.

6
Simulation of Stock Prices

The simulation of discrete events can help us in understanding of the data. In this chapter, we implement a simulation of the stock market by applying the random walk algorithm and present it with `D3.js` animation.

In this chapter we will cover:

- Financial time series
- Random walk simulation
- Monte Carlo methods
- Generating random numbers
- Implementation in `D3.js`

Financial time series

Financial time series analysis (FTSA) involves working with asset valuation over time, such as currency exchange or stock market prices. FTSA addresses a particular feature, the uncertainty in words of the famous American financier J. P. Morgan, when asked what the stock market will do, he replied:

"It will fluctuate."

The uncertainty of financial time series states that the volatility of a stock price cannot be directly observable. In fact, Louis Bachelier's Theory of Speculation (1900) postulated that prices fluctuate randomly.

In the following screenshot, we can see the time series of Apple Inc. Historical stock prices for the last 3 months. In fact, simple random processes can create a time series, which will closely resemble this real-time series. The random walk model is considered in FTSA as a statistical model for the movement of logged stock prices:

 We can download the Apple Inc. Historical Stock Prices from the Nasdaq website, http://www.nasdaq.com/symbol/aapl/historical#.UT1jrRypwOJ.

For a complete and broad reference, refer to the *Analysis of Financial Time Series* book, by *Ruey S. Tsay*. In this chapter, we will implement a random walk simulation in D3.js and in the following section, we will discuss random walk and Monte Carlo models.

Random walk simulation

Random walk is a simulation where a succession of random steps is used to represent an apparently random event. The interesting thing is that we can use this kind of simulation to see different outputs from a certain event by controlling the start point of the simulation and the probability distribution of the random steps. Similar to all the simulations, this simulation is just a simplified model of the original phenomena. However, a simulation may be useful and is a powerful visualization tool. There are different motions of random walks using different implementations. The most common are Brownian motion and binomial model.

In the following figure, we can see the simulated data from random walk model for logged stock prices:

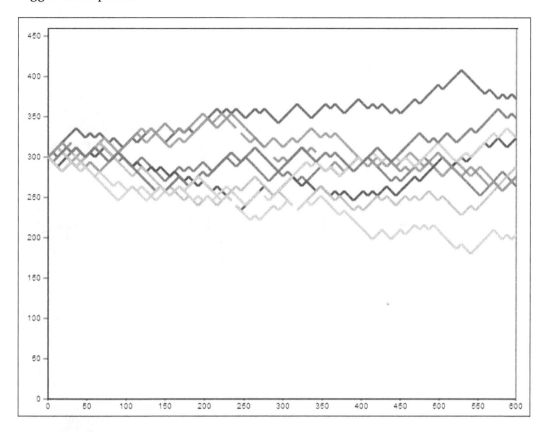

Brownian motion is a random walk model named after the physicist Robert Brown, who observed molecules moving and colliding with one another in random fashion. Brownian motion is usually used to model stock prices. According to the work of Robert C. Merton (Nobel laureate in Economics), the Brownian model of financial markets define that the stock prices evolve continuously in time and are driven by the Brownian motion processes. In this model we assumed a normal distribution of a returns period, this means that the probability of the random step does not vary over time and is independent of past steps.

Binomial model is a simple price model that is based on discrete steps, where the price of an asset can go up or down. If a price goes up then it is multiplied by an up-factor and on the other hand if the asset goes down then it is multiplied by a down-factor.

[For more information about the Brownian model of financial markets, please refer to the link `http://bit.ly/17WeyH7`.]

Monte Carlo methods

Random walk is a member of a family of random sampling algorithms, proposed by Stanislaw Ulam in 1940. Monte Carlo methods are mainly used when the event has uncertainty and deterministic boundaries (previous estimate of a range of limit values). These methods are especially good for optimization and numerical integration in biology, business, physics, and statistics.

Monte Carlo methods depend on the probability distribution of the random number generator to see different behaviors in the simulations. The most common distribution is the Gauss or normal distribution (see the following figure) but there are other distributions such as geometric or Poisson.

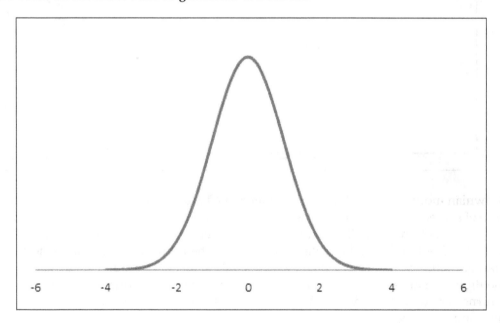

Generating random numbers

While getting truly random numbers is a difficult task, most of the Monte Carlo methods perform well with pseudo-random numbers and this makes it easier to re-run simulations based on a seed. Practically, all the modern programming languages include basic random sequences, at least good enough to make good simulations.

Python includes the random library. In the following code we can see the basic usage of the library:

- Importing the random library as rnd:

```
import random as rnd
```

- Getting a random float between 0 and 1:

```
>>>rnd.random()
0.254587458742659
```

- Getting a random number between 1 and 100:

```
>>>rnd.randint(1,100)
56
```

- Getting a random float between 10 and 100 using a uniform distribution:

```
>>>rnd.uniform(10,100)
15.2542689537156
```

 For a detailed list of methods of the random library, follow the link http://docs.python.org/3.2/library/random.html.

In the case of JavaScript, a more basic random function is included with the Math.random() function, however, for the purpose of this chapter, the random library will be enough.

In the following script, we can see a basic JavaScript code printing a random number between 0 and 100 in an HTML element with the ID label:

```
<script>
function  randFunction()
{
var x=document.getElementById("label")
x.innerHTML=Math.floor((Math.random()*100)+1);
}
</script>
```

Implementation in D3.js

In this chapter, we will create an animation in D3.js of a Brownian motion random walk simulation. In the simulation, we will control the delay of the animation, the starting point of the random walk, and the tendency of the up-down factor.

First, we need to create an HTML file named Simulation.html and we will run it from Python http.server. In order to run the animation, we just need to open a command terminal and run the following command:

```
>>python -m http.server 8000
```

Then, we just need to open a web browser and type http://localhost:8000 and select our HTML file, after that we can see the animation running.

Then, we need to import the D3 library either directly from the website or with a local copy of the d3.v3.min.js file.

```
<script type="text/javascript" src="http://d3js.org/d3.v3.min.js"></
script>
```

In the CSS, we specified the style for the axis line, the font family, size for the text, and the background color.

```
<style type="text/css">
body {
  background: #fff;
}
.axis text {
  font: 10px sans-serif;
}
.axis path,
.axis line {
  fill: none;
  stroke: #000;
}
</style>
```

 We can define the colors in CSS using a hexadecimal code such as #fff instead of the literal name white. We can find a color selector at http://www.w3schools.com/tags/ref_colorpicker.asp.

We need to define some variables that we will need in the animation such as the delay, the first line color, height, and width of the work area. The color variable will be randomly reassigned every time the time series reaches the edge of the canvas to start with a new color for the next time series.

```
var color = "rgb(0,76,153)";
var GRID = 6,
HEIGHT = 600,
WIDTH = 600,
delay = 50;
```

Now, we need to define the size of the new SVG's width and height (630 x 650 pixels, including extra space for the axis labels), which inserts a new <svg> element inside the <body> tag:

```
var svg = d3.select("body").append("svg:svg")
  .attr("width", WIDTH + 50)
  .attr("height", HEIGHT + 30)
 .append("g")
  .attr("transform", "translate(30,0)");
```

Now, we need to set the associated scale for the X axis and the Y axis, as well as the label's orientation:

```
var x = d3.scale.identity()
    .domain([0, WIDTH]);
var y = d3.scale.linear()
    .domain([0, HEIGHT])
    .range([HEIGHT, 0]);
var xAxis = d3.svg.axis()
    .scale(x)
    .orient("bottom")
    .tickSize(2, -HEIGHT);
var yAxis = d3.svg.axis()
    .scale(y)
    .orient("left")
    .tickSize(6, -WIDTH);
```

Append the axis to an SVG selection with the <g> element:

```
svg.append("g")
    .attr("class", "x axis")
    .attr("transform", "translate(0,600)")
    .call(xAxis);
  svg.append("g")
    .attr("class", "y axis")
  .attr("transform", "translate(0,0)")
    .call(yAxis);
```

We can define the D3 reference API documentation of SVG Axes at `https://github.com/mbostock/d3/wiki/SVG-Axes`.

Then, we will include a text label in the X axis (position `270`, `50`):

```
svg.append("text")
        .attr("x", 270 )
        .attr("y",  50 )
        .style("text-anchor", "middle")
        .text("Random Walk Simulation");
```

We will create a function called `randomWalk` to perform each step of the simulation. This will be a recursive function that includes the drawing of the line segments for each step of the random walk. And using the `Math.random()` function, we will decide if the walker goes up or down:

```
function randomWalk(x, y) {
var x_end, y_end = y + GRID;
if (Math.random() < 0.5) {
  x_end = x + GRID;
} else {
  x_end = x - GRID;
}
line = svg.select('line[x1="' + x + '"][x2="' + x_end + '"]'+
                '[y1="' + y + '"][y2="' + y_end + '"]');
```

Now, we need to add the new line segment to the svg element `svg:line` with a random color and 3 points of stroke width:

```
svg.append("svg:line")
    .attr("x1", y)
    .attr("y1", x)
    .attr("x2", y_end)
    .attr("y2", x_end)
    .style("stroke", color)
    .style("stroke-width", 3)
    .datum(0);
```

When the walker (y_end) reaches the end of the workspace, we need to pick a new color randomly, with the `Math.floor(Math.random()*254)` function in each of the RGB code and reset the control variables (y_end and x_end):

```
if (y_end >= HEIGHT) {
color = "rgb("+Math.floor(Math.random()*254)+",
            "+Math.floor(Math.random()*254)+",
```

```
                  "+Math.floor(Math.random()*254)+")"
  x_end = WIDTH / 2;
  y_end = 0;
}
```

With the `window.setTimeout` function, we will wait for 50 milliseconds to get the progressive effect of the animation and call the `randomWalk` function again.

```
  window.setTimeout(function() {
    randomWalk(x_end, y_end);
  }, delay);
}
```

Finally, we need to call the `randomWalk()` function to pass the starting point as a parameter in the Y axis of the animation.

```
randomWalk(WIDTH / 2, 0);
```

 All the code of this chapter can be found in the author's GitHub repository at `https://github.com/hmcuesta/PDA_Book/tree/master/Chapter6`.

In the following screenshot, we can see the result of the animation after 12 iterations in the screenshot labeled **1** and more iterations in the screenshot labeled **2**:

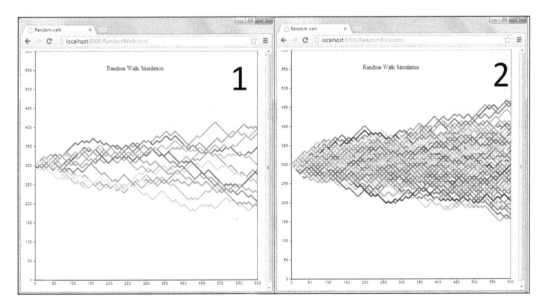

One interesting thing that we can observe is the normal distribution presented in the visualization. In the following figure, we can see the normal distribution of the random walk in a shaded area:

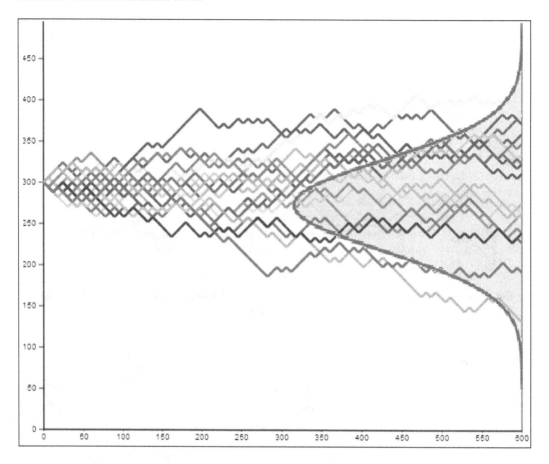

We can also try different start parameters to get different outputs such as changing the start point of the lines and the distribution of the random walk as we can see in the following screenshot:

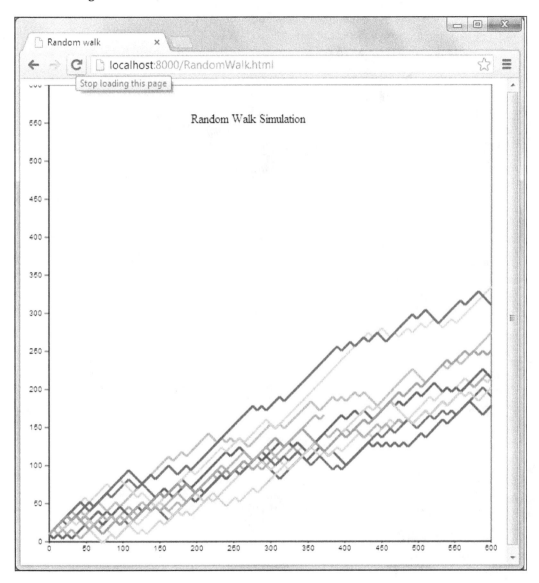

The complete code of the random walk simulator is listed as follows:

```html
<html>
  <head>
    <meta content="text/html;charset=utf-8">
    <title>Random walk</title>
    <script type="text/javascript"   src="http://d3js.org/d3.v3.min.
js">
</script>
    <style type="text/css">
 body {
   background: #fff;
}
.axis text {
  font: 10px sans-serif;
}
.axis path,
.axis line {
  fill: none;
  stroke: #000;
}
    </style>
</head>
<body>
<script>
var color = "rgb(0,76,153)";
var GRID = 6,
HEIGHT = 600,
WIDTH = 600,
delay = 50,
svg = d3.select("body").append("svg:svg")
  .attr("width", WIDTH + 50)
  .attr("height", HEIGHT + 30)
 .append("g")
  .attr("transform", "translate(30,0)");

var x = d3.scale.identity()
    .domain([0, WIDTH]);
var y = d3.scale.linear()
    .domain([0, HEIGHT])
    .range([HEIGHT, 0]);
var xAxis = d3.svg.axis()
    .scale(x)
    .orient("bottom")
```

```
    .tickSize(2, -HEIGHT);
var yAxis = d3.svg.axis()
    .scale(y)
    .orient("left")
    .tickSize(6, -WIDTH);

svg.append("g")
    .attr("class", "y axis")
  .attr("transform", "translate(0,0)")
    .call(yAxis);
svg.append("g")
    .attr("class", "x axis")
    .attr("transform", "translate(0,600)")
    .call(xAxis);
svg.append("text")
        .attr("x", 270 )
        .attr("y",  50 )
        .style("text-anchor", "middle")
        .text("Random Walk Simulation");
  randomWalk(WIDTH / 2, 0);
  function randomWalk (x, y) {
  var x_end, y_end = y + GRID;
  if (Math.random() < 0.5) {
    x_end = x + GRID;
  } else {
    x_end = x - GRID;
  } line = svg.select('line[x1="' + x + '"][x2="' + x_end + '"]'+
                  '[y1="' + y + '"][y2="' + y_end + '"]');
      svg.append("svg:line")
      .attr("x1", y)
      .attr("y1", x)
      .attr("x2", y_end)
      .attr("y2", x_end)
      .style("stroke", color)
      .style("stroke-width", 3)
      .datum(0);
    if (y_end >= HEIGHT) {
  color = "rgb("+Math.floor(Math.random()*254)+",
              "+Math.floor(Math.random()*254)+",
              "+Math.floor(Math.random()*254)+")"
    x_end = WIDTH / 2;
    y_end = 0;
  }
  window.setTimeout(function() {
```

```
        randomWalk(x_end, y_end);
    }, delay);
}
</script>
</body>
</html>
```

Summary

In this chapter, we explored the random walk simulation and how to communicate through animated visualizations. Simulation is an excellent way to see certain behavior of a phenomenon such as stock prices. The Monte Carlo methods are widely used to simulate phenomena when we don't have the means to reproduce an event because it is either dangerous or expensive such as epidemic outbreaks or stock prices. However, a simulation is always a simplified model of the real world. The goal of the simulation presented in this chapter is to show how we can get a basic but attractive web-based visualization with D3.

In the next chapter, we will learn the basic concepts of time series. Then, we will present a numeric prediction of gold prices using regression and classification techniques.

7

Predicting Gold Prices

In this chapter, you will be introduced to the basic concepts of **time series data** and **regression**. First, we distinguish some of the basic concepts such as trend, seasonality, and noise. Then we introduce the historic gold prices time series and also get an overview on how to perform a forecast using **kernel ridge regression**. Later, we present a regression using the smoothed time series as an input.

This chapter will cover:

- Working with the time series data
- The data – historical gold prices
- Nonlinear regression
- Kernel ridge regression
- Smoothing the gold prices time series
- Predicting in the smoothed time series
- Contrasting the predicted value

Working with the time series data

Time series is one of the most common ways to find data in the real world. A time series is defined as the changes of a variable through the time. **Time series analysis (TSA)** is widely used in economics, weather, and epidemiology. Working with time series needs to define some basic concepts of trend, seasonality, and noise.

In the following figure, found at `http://www.gold.org/investment/statistics/gold_price_chart/`, we can see the time series for gold price in US since July 2010.

Typically the easiest way to explore a time series is with a line chart. With the help of direct appreciation of the time series visualization, we can find anomalies and complex behavior in the data.

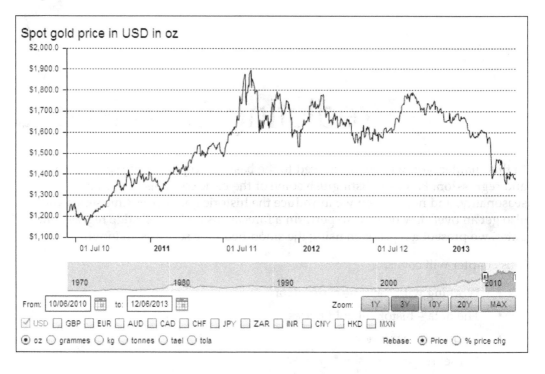

We have two kinds of time series; linear and nonlinear. In the following figure, we can see an example of each one. Plotting time series data is very similar to scatterplot or line chart, but the data points in X axis are times or dates:

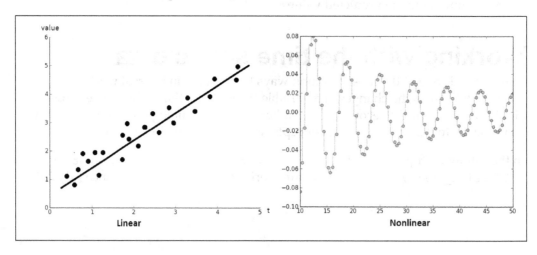

Components of a time series

In many cases, a time series is the sum of multiple components:

$$X_t = T_t + S_t + V_t$$

Observation = Trend + Seasonality + Variability

- **Trend (T)**: The behavior or slow motion in the time series through a large timeframe
- **Seasonality (S)**: The oscillatory motion in a year, for example, the flu season
- **Variability (V)**: The random variations around the previous components

In the following figure, we can see a time series with an evolutionary trend which doesn't follow a linear pattern and slowly evolves through the time:

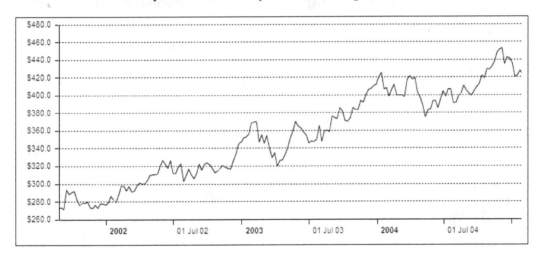

In this book, the visualization is driven with D3.js (web-based). However, it is important to have a fast visualization tool directly from the Python language. In this chapter, we will use matplotlib as a standalone visualization tool. In the following code, we can see an example of how to use matplotlib to visualize a line chart.

First, we need to import the library and assign an alias plt:

```
import matplotlib.pyplot as plt
```

Then, using the numpy library, we will create a synthetic data with the linspace and cos methods for the x and y data respectively:

```
import numpy as np
x = np.linspace(10, 100, 500)
y = np.cos(x)/x
```

Now, we prepare the visualization with the step function and present the visualization in a new window using the show function:

```
plt.step(x, y)
plt.show()
```

 You can find more information about matplotlib at http://matplotlib.org/.

Finally, the following screenshot displays the visualization window with the result.

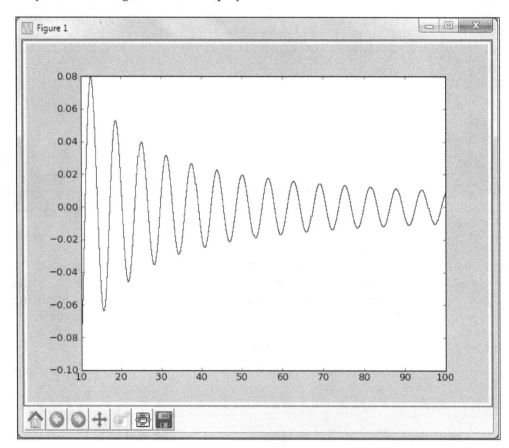

As we can see in the preceding screenshot, the visualization window provides us with some tools such as pan axes, zoom, and save, that help us to prepare and export the visualization in a `.png` image format. We can also navigate through the changes or go back to the original view.

Smoothing the time series

When we work with real-world data, we may often find noise, which is defined as pseudo-random fluctuations in values that don't belong to the observation data. In order to avoid or reduce this noise, we can use different approaches such as increasing the amount of data by the interpolation of new values where the series is sparse. However, in many cases this is not an option. Another approach is smoothing the series, typically using the **average** or **exponential** methods. The average method helps us to smooth the series by replacing each element in the series with either simple or weighted average of the data around it. We will define a **Smoothing Window** to the interval of possible values which control the smoothness of the result. The main disadvantage of using the moving averages approach is, if we have outliers or abrupt jumps in the original time series, the result may be inaccurate and can produce jagged curves.

In this chapter, we will implement a different approach using convolution (moving averages filter) of a scaled window with the signal. This approach is taken from **Digital Signal Processing (DSP)**. In this case, we use a time series (signal) and we will apply a filter, getting a new time series as a result. In the following code, we can see an example of how to smooth a time series. For this example, we will use the log of USA/CAD historical exchange rates from March 2008 to March 2013 with 260 records.

 The historical exchange rates can be downloaded from http://www.oanda.com/currency/historical-rates/.

The first seven records of the CSV file (`ExchangeRate.csv`) look as follows:

```
date,usd
3/10/2013,1.028
3/3/2013,1.0254
2/24/2013,1.014
2/17/2013,1.0035
2/10/2013,0.9979
2/3/2013,1.0023
1/27/2013,0.9973
...
```

First, we need to import all the required libraries, see *Appendix, Setting Up the Infrastructure,* for complete installation instructions for numpy and scipy libraries:

```
import dateutil.parser as dparser
import matplotlib.pyplot as plt
import numpy as np
from pylab import *
```

Now, we will create the smooth function, setting the original time series and the windows length as parameters. In this implementation, we use the numpy implementation of the **Hamming** window (np.hamming); however, we can use other kinds of window such as **Flat**, **Hanning**, **Bartlett**, and **Blackman**.

For complete reference of the window functions supported by numpy, please refer to http://docs.scipy.org/doc/numpy/reference/routines.window.html.

```
def smooth(x,window_len):
  s=np.r_[2*x[0]-x[window_len-1::-1],
    x,2*x[-1]-x[-1:-window_len:-1]]
  w = np.hamming(window_len)
    y=np.convolve(w/w.sum(),s,mode='same')
    return y[window_len:-window_len+1].
```

The method presented in this chapter is based on the signal smoothing from scipy reference documentation and can be found at http://wiki.scipy.org/Cookbook/SignalSmooth.

Then, we need to obtain the labels for the X axis, using the numpy genfromtxt function to get the first column in the CSV file and applying a converter function dparser.parse to parse the date data:

```
x = np.genfromtxt("ExchangeRate.csv",
  dtype='object',
  delimiter=',',
  skip_header=1,
  usecols=(0),
  converters = {0: dparser.parse})
```

Now, we need to obtain the original time series from the `ExchangeRate.csv` file:

```
originalTS = np.genfromtxt("ExchangeRate.csv",
  skip_header=1,
  dtype=None,
  delimiter=',',
  usecols=(1))
```

Then, we apply the `smooth` method and store the result in the `smoothedTS` list:

```
smoothedTS = smooth(originalTS, len(originalTS))
```

Finally, we plot the two series using `pyplot`:

```
plt.step(x, originalTS, 'co')
plt.step(x, smoothedTS)
plt.show()
```

In the following image, we can see the original (dotted line) and the smoothed (line) series. We can observe that in the visualization that in the smoothed series we cut out the irregular roughness to see a clearer signal. Smoothing doesn't provide us with a model per se. However, it can be the first step to describe multiple components of the time series. When we work with epidemiological data, we can smooth out the seasonality so that we can identify the trend (See *Chapter 10, Working with Social Graphs*).

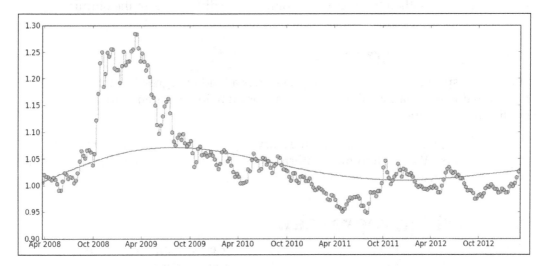

The data – historical gold prices

Regression analysis is a statistical tool for understanding the relationship between variables. In this chapter, we will implement a nonlinear regression to predict the gold price based on the historic gold prices. For this example, we will use the historical gold prices from January 2003 to May 2013 in a monthly range, obtained from www.gold.org. Finally, we will forecast the gold price for June 2013 and will contrast it with the real price from an independent source. The complete datasets (since December 1978) can be found at http://gold.org/download/value/stats/statistics/xls/gold_prices.xls.

The first seven records of the CSV file (gold.csv) look as follows:

```
date,price
1/31/2003,367.5
2/28/2003,347.5
3/31/2003,334.9
4/30/2003,336.8
5/30/2003,361.4
6/30/2003,346.0
7/31/2003,354.8
```

In this example, we will implement a Kernel ridge regression with the original time series and the smoothed time series, to compare the differences in the output.

Nonlinear regression

Statistically speaking the nonlinear regression is a kind of regression analysis for estimating the relationships between one or more independent variables in a nonlinear combination.

In this chapter, we will use the Python library mlpy and its Kernel ridge regression implementation. We can find more information about nonlinear regression methods at http://mlpy.sourceforge.net/docs/3.3/nonlin_regr.html.

Kernel ridge regression

The most basic algorithm that can be kernelized is **Kernel ridge regression (KRR)**. It is similar to an **SVM (Support Vector Machines)** (see *Chapter 8, Working with Support Vector Machines*) but the solution depends on all the training samples and not on the subset of support vectors. KRR works well with few training sets for classification and regression. In this chapter, we will focus on its implementation using mlpy rather than all the linear algebra involved. See *Appendix, Setting Up the Infrastructure*, for complete installation instructions for mlpy library.

First, we need to import the `numpy`, `mlpy`, and `matplotlib` libraries:

```
import numpy as np
import mlpy
from mlpy import KernelRidge
import matplotlib.pyplot as plt
```

Now, we define the seed for the random number generation:

np.random.seed(10)

Then we need to load the historical gold prices from the `Gold.csv` file and store them in `targetValues`:

```
targetValues = np.genfromtxt("Gold.csv",
  skip_header=1,
  dtype=None,
  delimiter=',',
  usecols=(1))
```

Next we will create a new array with 125 training points, one for each record of the `targetValues` representing the monthly gold price from Jan 2003 to May 2013:

```
trainingPoints = np.arange(125).reshape(-1, 1)
```

Then, we will create other array with 126 test points representing the original 125 points in `targetValues` and including an extra point for our predicted value for Jun 2013:

```
testPoints = np.arange(126).reshape(-1, 1)
```

Now, we create the training kernel matrix (`knl`) and testing kernel matrix (`knlTest`). Kernel ridge regression (KRR) will randomly split the data into subsets of same size, then process an independent KRR estimator for each subset. Finally, we average the local solutions into a global predictor:

```
knl = mlpy.kernel_gaussian(trainingPoints, trainingPoints,
  sigma=1)
knlTest = mlpy.kernel_gaussian(testPoints, trainingPoints,
  sigma=1)
```

Then, we instance the `mlpy.KernelRidge` class in the `knlRidge` object:

```
knlRidge = KernelRidge(lmb=0.01, kernel=None)
```

The `learn` method will compute the regression coefficients, using the training kernel matrix and the target values as a parameters:

```
knlRidge.learn(knl, targetValues)
```

The `pred` method computes the predicted response, using the testing kernel matrix as an input:

```
resultPoints = knlRidge.pred(knlTest)
```

Finally, we plot the two time series of target values and result points:

```
fig = plt.figure(1)
plot1 = plt.plot(trainingPoints, targetValues, 'o')
plot2 = plt.plot(testPoints, resultPoints)
plt.show()
```

In the following figure, we can observe the points which represents the target values (the known values) and the line that represent the result points (result from the `pred` method). We may observe the last segment of the line which is the predicted value for June 2013:

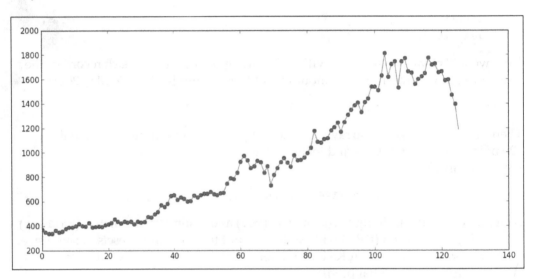

In the following screenshot, we can observe the resulted points from the `knlRidge.pred()` method and the last value (1186.16129538) is the predicted value for June 2013:

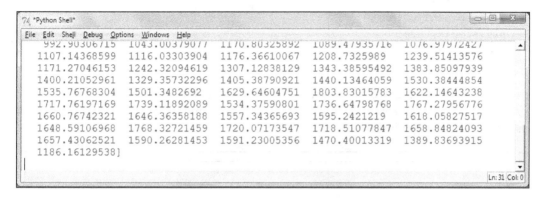

All the codes and datasets of this chapter may be found in the author's GitHub repository at `https://github.com/hmcuesta/PDA_Book/tree/master/Chapter7`.

Smoothing the gold prices time series

As we can see the gold prices time series is noisy and it's hard to spot a trend or patterns with a direct appreciation. So to make it easier, we may smooth the time series. In the following code, we smooth the gold prices time series (see *Smoothing time series* section in this chapter for a detailed explanation):

```
import matplotlib.pyplot as plt
import numpy as np
import dateutil.parser as dparser
from pylab import *
def smooth(x,window_len):
    s=np.r_[2*x[0]-x[window_len-1::-1],x,2*x[-1]-x[-1:-window_len:-1]]
    w = np.hamming(window_len)
    y=np.convolve(w/w.sum(),s,mode='same')
    return y[window_len:-window_len+1]
x = np.genfromtxt("Gold.csv",
    dtype='object',
    delimiter=',',
    skip_header=1,
    usecols=(0),
    converters = {0: dparser.parse})
y = np.genfromtxt("Gold.csv",
    skip_header=1,
```

```
        dtype=None,
        delimiter=',',
        usecols=(1))
y2 = smooth(y, len(y))
plt.step(x, y2)
plt.step(x, y, 'co')
plt.show()
```

In the following figure, we can observe the time series of the historical gold prices (the dotted line) and we can see the smoothed time series (the line) using the hamming window:

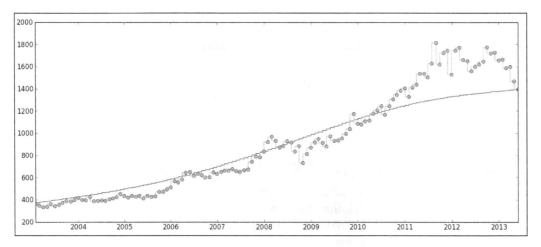

Predicting in the smoothed time series

Finally, we put everything together and implement the Kernel ridge Regression to the smoothed gold prices time series. We can find the complete code of the KRR as follows:

```
import matplotlib.pyplot as plt
import numpy as np
import dateutil.parser as dparser
from pylab import *
import mlpy
def smooth(x,window_len):
    s=np.r_[2*x[0]-x[window_len-1::-1],
        x,2*x[-1]-x[-1:-window_len:-1]]
    w = np.hamming(window_len)
```

```
    y=np.convolve(w/w.sum(),s,mode='same')
    return y[window_len:-window_len+1]
y = np.genfromtxt("Gold.csv",
    skip_header=1,
    dtype=None,
    delimiter=',',
    usecols=(1))
targetValues = smooth(y, len(y))
np.random.seed(10)
trainingPoints = np.arange(125).reshape(-1, 1)
testPoints = np.arange(126).reshape(-1, 1)
knl = mlpy.kernel_gaussian(trainingPoints,
    trainingPoints, sigma=1)
knlTest = mlpy.kernel_gaussian(testPoints,
    trainingPoints, sigma=1)
knlRidge = mlpy.KernelRidge(lmb=0.01, kernel=None)
knlRidge.learn(knl, targetValues)
resultPoints = knlRidge.pred(knlTest)

plt.step(trainingPoints, targetValues, 'o')
plt.step(testPoints, resultPoints)
plt.show()
```

In the following figure, we can observe the dotted line which represents the smoothed time series of the historical gold prices, and the line that represents the prediction for the gold price in June 2013:

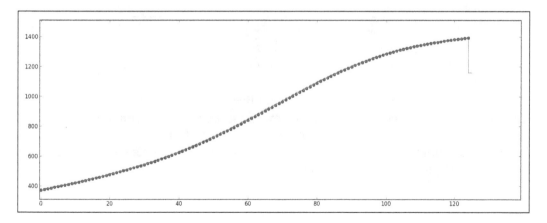

In the following screenshot, we can see the predicted values for the smoothed time series. This time we can observe that the values are much lower than the original predictions:

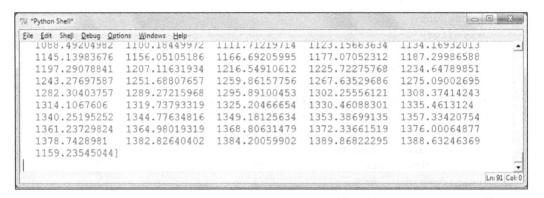

Contrasting the predicted value

Finally, we will look for an external source to see if our prediction is realistic. In the following figure, we may observe a graph from *The Guardian/Thomson Reuters* for June 2013. The gold price fluctuated between 1180.0 and 1210.0 with an official average of 192.0 for the month. Our prediction for the Kernel ridge regression with complete data is 1186.0, which is not bad at all. We can see the complete numbers in the following table:

Source	June 2013
The Guardian/Thomson Reuters (external Source)	1192.0
Kernel ridge regression with complete data (predictive model)	1186.161295
Kernel ridge regression with smoothed data (predictive model)	1159.23545044

A good practice when we want to build a predictive model is to try different approaches for the same problem. If we develop more than one model, we may compare testing results against each other and select the best model. For this particular example, the value predicted using the complete data is more accurate than the value predicted using the smoothed data.

In words of the mathematician named George E. P. Box:

"All models are wrong, but some are useful"

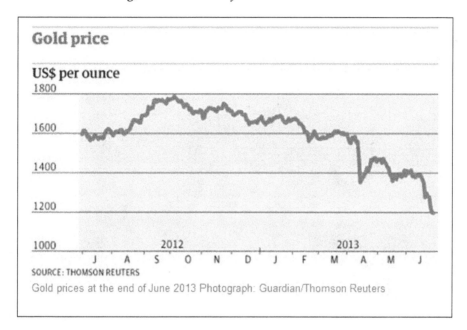

SOURCE: THOMSON REUTERS

Gold prices at the end of June 2013 Photograph: Guardian/Thomson Reuters

 For the complete information about the article *Stock markets and gold suffer a June to forget*, please refer to `http://www.theguardian.com/business/2013/jun/28/stock-markets-gold-june`.

Summary

In this chapter, we explored the nature of time series, describing their components and implementing signal processing to smooth the time series. Then, we introduced the Kernel ridge regression (KRR) implemented in the `mlpy` library. Finally we presented two implementations of the KRR; one with the complete data and the other with the smoothed data, to predict the monthly gold price in June 2013 and we found that for this case the prediction with the complete data was more accurate.

In the next chapter, we will learn how to perform a dimensionality reduction and how to implement a support vector machine (SVM) with a multivariate dataset.

8
Working with Support Vector Machines

The **support vector machine (SVM)** is a powerful classification technique. In this chapter, we will provide the reader with an easy way to get acceptable results using SVM. We will perform dimensionality reduction of the dataset and we will produce a model for classification.

The theoretical foundation of SVM lies in the work of Vladimir Vapnik and the theory of statistical learning developed in the 1970s. The SVMs are highly used in pattern recognition of Time Series, Bioinformatics, Natural Language Processing, and Computer Vision.

In this chapter, we will use the `mlpy` implementation of LIBSVM, which is a widely used library for SVM with several interfaces and extensions for languages such as Java, Python, MATLAB, R, CUDA, C#, and Weka. For more information about LIBSVM visit the following link:

`http://www.csie.ntu.edu.tw/~cjlin/libsvm/`

In this chapter we will cover:

- Understanding the multivariate dataset
- Dimensionality Reduction
    - **Linear Discriminant Analysis (LDA)**
    - **Principal Component Analysis (PCA)**

- Getting started with support vector machines
  - ° Kernel functions
  - ° Double spiral problem
  - ° SVM implementation using mlpy

Understanding the multivariate dataset

A multivariate dataset is defined as a set of multiple observations (attributes) associated with different aspects of a phenomenon. In this chapter, we will use a multivariate dataset result of a chemical analysis of wines that grew in three different cultivars from the same area in Italy. The Wine dataset is available in the UC Irvine Machine Learning Repository and can be freely downloaded from the following link:

http://archive.ics.uci.edu/ml/datasets/Wine

The dataset includes 13 features with no missing data and all the features are numerical or real values.

The complete list of features is listed as follows:

- Alcohol
- Malic acid
- Ash
- Alkalinity of ash
- Magnesium
- Total phenols
- Flavonoids
- Nonflavonoid phenols
- Proanthocyanins
- Color intensity
- Hue
- OD280/OD315 of diluted wines
- Proline

The dataset has 178 records from three different classes. The distribution is seen in the following figure corresponding to 59 for class 1, 71 for class 2, and 48 for class 3:

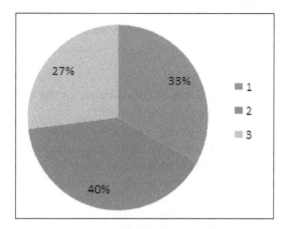

The first five records of the dataset will look as follows:

```
1,14.23,1.71,2.43,15.6,127,2.8,3.06,.28,2.29,5.64,1.04,3.92,1065
1,13.2,1.78,2.14,11.2,100,2.65,2.76,.26,1.28,4.38,1.05,3.4,1050
1,13.16,2.36,2.67,18.6,101,2.8,3.24,.3,2.81,5.68,1.03,3.17,1185
1,14.37,1.95,2.5,16.8,113,3.85,3.49,.24,2.18,7.8,.86,3.45,1480
1,13.24,2.59,2.87,21,118,2.8,2.69,.39,1.82,4.32,1.04,2.93,735
```

In the following code snippet, we will plot two of the features from the dataset at a time. In this example we will plot Alcohol and Malic acid attributes. However, to visualize all the possible features' combination we will need the binomial coefficient of the number of features. In this case, 13 features are equal to 78 different combinations. Due to this, it is mandatory to perform dimensionality reduction. Perform the following steps:

```
import matplotlib
import matplotlib.pyplot as plt
```

1. Firstly, we will obtain the data from the dataset into a matrix of the features and a list of the categories associated with each record with the getData function.

```
def getData():
    lists = [line.strip().split(",") for line in open
  ('wine.data', 'r').readlines()]
    return [list( l[1:14]) for l in lists], [l[0] for l in lists]

matrix, labels = getData()
```

```
xaxis1 = [] ; yaxis1 = []
xaxis2 = [] ; yaxis2 = []
xaxis3 = [] ; yaxis3 = []
```

2. Then, we will select the two features to visualize in the variables x and y.

```
x = 0 #Alcohol
y = 1 #Malic Acid
```

3. Next, we will generate the sets of coordinates for the two attributes (x,y) of the three categories (classes) of the dataset.

```
for n, elem in enumerate(matrix):
    if int(labels[n]) == 1:
        xaxis1.append(matrix[n][x])
        yaxis1.append(matrix[n][y])
    elif int(labels[n]) == 2:
        xaxis2.append(matrix[n][x])
        yaxis2.append(matrix[n][y])
    elif int(labels[n]) == 3:
        xaxis3.append(matrix[n][x])
        yaxis3.append(matrix[n][y])
```

4. Finally, we will plot the three categories (classes) into a scatter plot.

```
fig = plt.figure()
ax = fig.add_subplot(111)
type1 = ax.scatter(xaxis1, yaxis1, s=50, c='white')
type2 = ax.scatter(xaxis2, yaxis2, s=50, c='red')
type3 = ax.scatter(xaxis3, yaxis3, s=50, c='darkred')

ax.set_title('Wine Features', fontsize=14)
ax.set_xlabel('X axis')
ax.set_ylabel('Y axis')
ax.legend([type1, type2, type3], ["Class 1", "Class 2", "Class
3"], loc=1)
ax.grid(True,linestyle='-',color='0.80')

plt.show()
```

In the following screenshot, we can observe the plot result of the features, such as Alcohol and Malic acid, for the three classes:

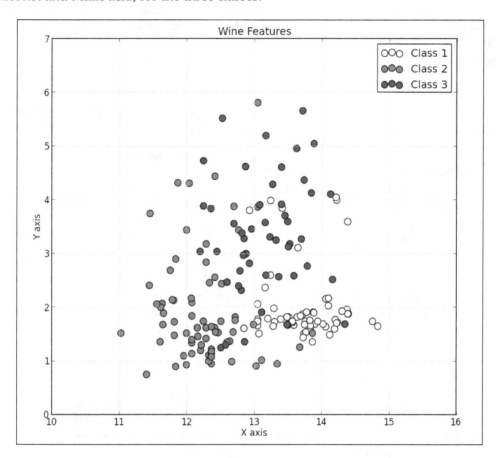

 We may also use a scatter plot matrix to visualize all the features in the dataset. However, this is computationally expensive and the number of subplots will depend on the binomial coefficient of the number of features. Refer to *Chapter 14, Online Data Analysis with IPython and Wakari*, to see how to plot the multidimensional datasets with scatterplot matrix and RadViz.

Dimensionality reduction

The dimensionality of a model is the number of independent attributes in the dataset. In order to reduce the complexity of the model we need to reduce the dimensionality without sacrificing accuracy. When we work in complex multidimensional data, we need to select the features that can improve the accuracy of the technique that we are using. Sometimes, we don't know if the variables are independent or if they share some kind of relationship. We need some criteria to find the best features and reduce the number of variables under consideration. In order to address these problems, we will perform three techniques: feature selection, feature extraction, and dimension reduction.

- **Feature selection**: We will select a subset of features in order to get better training times or improve the model accuracy. In data analysis, finding the best features for our problem is often guided by intuition and we don't know the real value of a variable until we test it. However, we may use metrics, such as correlation and mutual information that help by providing us with distance between features. The correlation coefficient is a measure of how strong is the relationship between two variables, and mutual information refers to a measure of how much one variable tells about another.

- **Feature extraction**: It is a special form of dimensionality reduction technique, performed by a transform in a high-dimensional space (multivariate dataset) to get a space of a fewer dimensions (the ones more informative). Two of the classical algorithms in the field are PCA or **multidimensional scaling** (**MDS**). Feature extraction is widely used in image processing, computer vision, and data mining.

- **Dimension reduction**: When we work with high-dimensional data there are various phenomena that may affect the result of our analysis and this is known as a "curse of dimensionality". In order to avoid these problems we will apply a preprocessing step using PCA or LDA.

 We may find more information about "curse of dimensionality" at the link http://bit.ly/7xJNzm.

Linear Discriminant Analysis

The LDA is a statistical method used to find linear combination of features, which can be used as a linear classifier. LDA is often used as a dimensionality reduction step before a complex classification. The main difference between LDA and PCA is that PCA does feature extraction and LDA performs classification. The `mlpy` implementation of LDA may be found at the following link:

http://bit.ly/19xyq3H

Principal Component Analysis

The PCA is the most used dimensionality reduction algorithm. PCA is an algorithm used to find a subset of features lineally uncorrelated known as principal components. PCA may be used in **exploratory data analysis** (**EDA**) through visual methods to find the most important characteristics in a dataset. This time we will implement a feature, selection, and PCA to the Wine dataset. In the following code snippet, we present the basic implementation of PCA in `mlpy`.

1. Firstly, we need to import the `numpy`, `mlpy`, and `matplotlib` modules. Refer to *Appendix, Setting Up the Infrastructure*, for installation instructions of Python Modules.

   ```
   import numpy as np
   import mlpy
   import matplotlib.pyplot as plt
   import matplotlib.cm as cm
   ```

2. Next, we will open the `wine.data` file using the `numpy` function `loadtxt`.

   ```
   wine = np.loadtxt('wine.data', delimiter=',')
   ```

3. Then, we will define the features in this case we will select the features 2 (malic acid), 3 (ash), and 4 (alkalinity of ash) for axis X and the class (which is the feature 0) for axis Y.

   ```
   x, y = wine[:, 2:5], wine[:, 0].astype(np.int)
   print(x.shape)
   print(y.shape)
   ```

   In this case the `x.shape` and `y.shape` will look as follows:

   ```
   >>> (178,3)
   >>> (178,)
   ```

4. Now, we will create a new instance of PCA and we need to train the algorithm with selected features in x using the function `learn`.

   ```
   pca = mlpy.PCA()
   pca.learn(x)
   ```

5. Then, we will apply the dimensionality reduction to the features in the variable x and turn it into two-dimensional subspace with the parameter `k = 2`.

   ```
   z = pca.transform(x, k=2)
   ```

6. The result of the transformation will be stored in the variable z and its shape will look as follows:

   ```
   print(z.shape)
   >>> (178,2)
   ```

7. Finally, we will use `matplotlib` to visualize the scatter plot of the new two-dimensional subspace of the PCA store in `z`.

```
fig1 = plt.figure(1)
title = plt.title("PCA on wine dataset")
plot = plt.scatter(z[:, 0], z[:, 1], c=y, s=90, cmap=cm.Reds)
labx = plt.xlabel("First component")
laby = plt.ylabel("Second component")
plt.show()
```

In the following screenshot, we can observe the scatter plot of the PCA result using green, blue, and red to highlight each class.

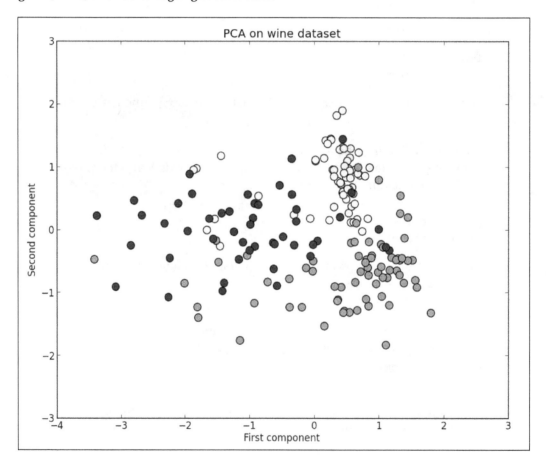

We can try different selections of the features and see what the result is. When the distribution of the data is highly dense, we will prefer to select options less attributes or mix some of them using proportions, or means. In the following screenshot, we can see the same implementation using more attributes Alcohol, Malic acid, Ash, Alkalinity of ash, Magnesium, and Total phenols. Due to this, we can see a different distribution of the points in the scatter plot.

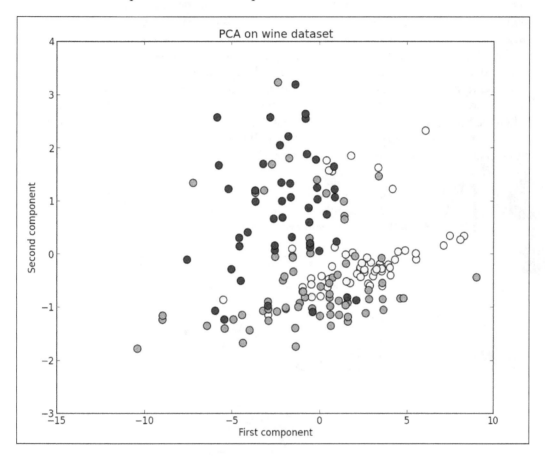

Getting started with support vector machine

The SVM is a supervised classification method based in a kernel geometrical construction as is shown in the following figure. SVM can be applied either for classification or regression. SVM will look for the best decision boundary that split the points into the class that they belong. To accomplish this SVM, we will look for the largest margin (space that is free of training samples parallel to the decision boundary). In the following figure, we can see the margin as the space between the dividing line and dotted lines. SVM will always look for a global solution due to the algorithm only care about the vectors close to the decision boundary. Those points in the edge of the margin are the support vectors. However, this is only for two-dimensional spaces, when we have high-dimensional spaces the decision boundaries turn into hyperplane (maximum decision margin) and the SVMs will look for the maximum-margin hyperplanes. In this chapter we will only work with two dimensional spaces.

> We may find more information, and reference of support vector machines and other kernel based techniques, at `http://www.support-vector-machines.org/`.

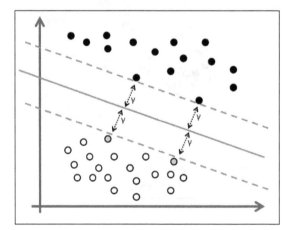

Kernel functions

The linear SVM has two main restrictions. First the resulted classifier will be linear, and second we need a dataset that can be split linearly. However, in the real world many data problems are not linear models. Due to this, we may want to try different kinds of kernels. SVMs support many different kinds of kernels but the most common are as follows:

- **Polynomial**: $(<gamma*uT*v>+coef0)^{degree}$ in `mlpy` defined as `kernel_type = "poly"`.

- **Gaussian**: $\exp\left(\dfrac{-(u-v)^2}{gamma}\right)$ in `mlpy` defined as `kernel_type = "rbf"`.

- **Sigmoidal**: $\tanh(gamma*uT*v+coef0)$ in `mlpy` defined as `kernel_type = "sigmoid"`.

- **Inverse multi-quadratic**: $\dfrac{1}{\sqrt{(u-gamma)^2+coef0^2}}$ not supported in `mlpy`.

Double spiral problem

The double spiral problem is a complex artificial problem that tries to distinguish between two classes with spiral shape. This problem is particularly hard for the classic classifiers due to its hard mix of values. The dataset are two classes in a spiral with 3 turns and 194 points. In the following screenshot we use SVMs with a Gaussian kernel, and we test the algorithm with different values for gamma. The gamma attribute defines the distance of a single training sample. If the gamma value is low, the attribute is farther and if the value is high, the attribute is nearer. The algorithm gives better solutions, as we increment gamma up to the value 100.

For the following screenshot, we use the code and dataset from the `mlpy` reference documentation and we can find it from the link http://bit.ly/18SjaiC.

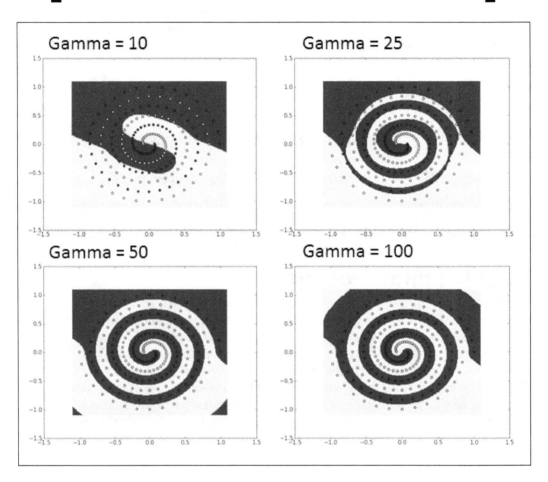

SVM implemented on mlpy

In the following code, we provide a simple implementation of the SVM algorithm with `mlpy`, which implements the LIBSVM library. In this case we will use a linear kernel, assuming that z is the two-dimensional space result of the PCA dimensionality reduction.

1. Firstly, we need to create a new instance of `svm` and define the kernel type as `linear`.

```
svm = mlpy.LibSvm(kernel_type='linear')
```

2. Then, we will train the algorithm with the function `learn` using as parameters the two dimensional space in the variable `z` and the class that belongs stored in the variable `y`.

```
svm.learn(z, y)
```

3. Now, we need to create a grid, where SVM will perform the predictions in order to visualize the result. We will use the `numpy` functions (such as `meshgrid` and `arange`) to create the matrix and then with revel function turn the matrices in a list of values for the predictor.

```
xmin, xmax = z[:,0].min()-0.1, z[:,0].max()+0.1
ymin, ymax = z[:,1].min()-0.1, z[:,1].max()+0.1
xx, yy = np.meshgrid(np.arange(xmin, xmax, 0.01),
          np.arange(ymin, ymax, 0.01))
grid = np.c_[xx.ravel(), yy.ravel()]
```

4. Next, with the `pred` function, we will return the prediction for each point in grid.

```
result = svm.pred(grid)
```

5. Finally, we will visualize the predictions in a scatter plot.

```
fig2 = plt.figure(2)
title = plt.title("SVM (linear kernel) on PCA")
plot1 = plt.pcolormesh(xx, yy, result.reshape(xx.shape), cmap=cm.
Greys_r)
plot2 = plt.scatter(z[:, 0], z[:, 1], c=y, s=90, cmap=cm.Reds)
labx = plt.xlabel("First component")
laby = plt.ylabel("Second component")
limx = plt.xlim(xmin, xmax)
limy = plt.ylim(ymin, ymax)
plt.show()
```

In the following screenshot, we can see the result of the plot for the SVM using a linear kernel.

We can observe a clear separation of the three classes. We may also see that the solution does not depend on all points; instead the separation will depend only on those points that are close to the decision boundary.

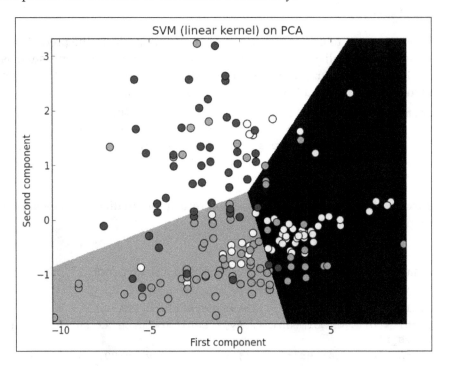

We execute SVM using more attributes such as Alcohol, Malic acid, Ash, Alkalinity of ash, Magnesium, and Total phenols in the following screenshot. Due to this the plot has different decision boundaries. However, if the SVM can't find a linear separation, the code will get into an infinite loop.

In the following screenshot we can see the result of the SVM implementing a Gaussian kernel and we can observe non-linear boundaries. The instruction that we need to update to get this result is given as follows:

```
svm = mlpy.LibSvm(kernel_type='rbf' gamma = 20)
```

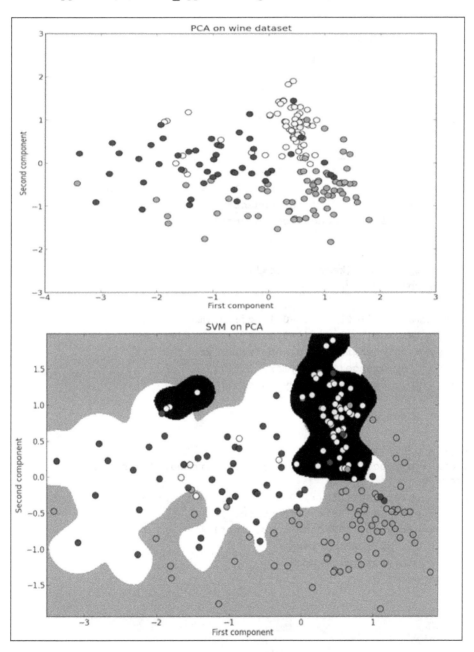

The complete code for the Wine classifier using PCA and SVM is listed as follows:

```
import numpy as np
import mlpy
import matplotlib.pyplot as plt
import matplotlib.cm as cm

wine = np.loadtxt('wine.data', delimiter=',')
x, y = wine[:, 2:5], wine[:, 0].astype(np.int)

pca = mlpy.PCA()
pca.learn(x)
z = pca.transform(x, k=2)

fig1 = plt.figure(1)
title = plt.title("PCA on wine dataset")
plot = plt.scatter(z[:, 0], z[:, 1], c=y, s=90, cmap=cm.Reds)
labx = plt.xlabel("First component")
laby = plt.ylabel("Second component")
plt.show()

svm = mlpy.LibSvm(kernel_type='linear')
svm.learn(z, y)

xmin, xmax = z[:,0].min()-0.1, z[:,0].max()+0.1
ymin, ymax = z[:,1].min()-0.1, z[:,1].max()+0.1
xx, yy = np.meshgrid(np.arange(xmin, xmax, 0.01),
        np.arange(ymin, ymax, 0.01))
grid = np.c_[xx.ravel(), yy.ravel()]

result = svm.pred(grid)

fig2 = plt.figure(2)
plot1 = plt.pcolormesh(xx, yy, result.reshape(xx.shape), cmap=cm.
Greys_r)
plot2 = plt.scatter(z[:, 0], z[:, 1], c=y, s=90, cmap=cm.Reds)
labx = plt.xlabel("First component")
laby = plt.ylabel("Second component")
limx = plt.xlim(xmin, xmax)
limy = plt.ylim(ymin, ymax)
plt.show()
```

All the codes and datasets of this chapter can be found in the author's GitHub repository at the link `https://github.com/hmcuesta/PDA_Book/tree/master/Chapter8`.

Summary

In this chapter, we get into dimensionality reduction and linear classification using SVM. In our example, we create a simple but powerful SVM classifier and we learn how to perform dimensionality reduction using PCA implemented in Python with `mlpy`. Finally, we present how to use non-linear kernels, such as Gaussian or Polynomial. The work in this chapter is just an introduction to the SVM algorithm with only two dimensions, the results can be improved with a multidimensional approach.

In the next chapter, we will learn how to model an epidemiological event (an infectious disease), and how to simulate an outbreak with cellular automation implemented in `D3.js`.

9
Modeling Infectious Disease with Cellular Automata

One of the goals of data analysis is to understand the system we are studying and modeling is the natural way to understand a real-world phenomenon. A model is always a simplified version of the real thing. However, through modeling and simulation we can try scenarios that are hard to reproduce, or are expensive, or dangerous. We can then perform analysis, define thresholds, and provide the information needed to make decisions. In this chapter, we will model an infectious disease outbreak through cellular automaton simulation implemented in JavaScript using D3.js. Finally, we will contrast the results of the simulation with the classical ordinary differential equations.

In this chapter, we will cover:

- Introduction to epidemiology
    - The epidemiology triangle
- The epidemic models:
    - The SIR model
    - Solving ordinary differential equation for the SIR model with SciPy
    - The SIRS model
- Modeling with cellular automata:
    - Cell, state, grid, and neighborhood
    - Global stochastic contact model
- Simulation of the SIRS model in CA with D3.js

Introduction to epidemiology

We can define epidemiology as the study of the determinants and distribution of health-related states. We will study how a pathogen is spread into a population such as common flu or influenza AH1N1. This is particularly important because an outbreak can cause severe human and economic loss. In the following screenshot we can see the interface of **Google Flu Trends (GFT)**, which uses an aggregated Google search data to estimate flu activity:

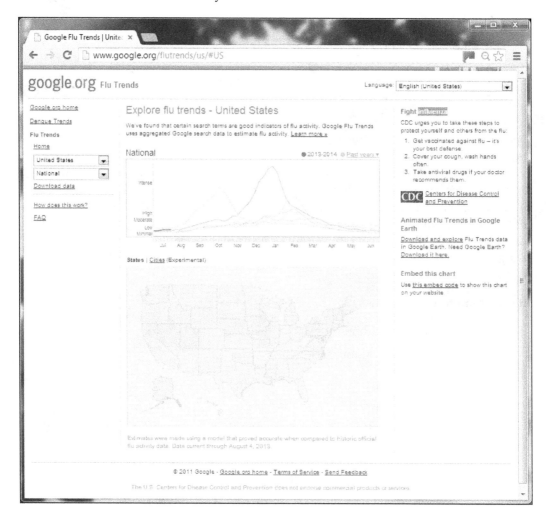

We can find the GFT data freely available from its website. With this time series we can implement statistical methods for descriptive epidemiology or causal inference. The GFT has been extensively compared with the seasonal influenza data of the **Center for Disease Control (CDC)**, which is obtained with typical surveys and medical reports, providing similar results.

We can find the Google Flu Trends data freely available at `http://www.google.org/flutrends/us/#US`.

And the **seasonal influenza (Flu)** data can be found at `http://www.cdc.gov/flu/`.

The epidemiology triangle

In the following screenshot we can see the epidemiologic triangle, which presents all the elements involved in an epidemic outbreak. We can see the **Agent**, which is the infectious pathogen. The **Host**—who is the human susceptible to catch the disease—will be highly related with its behavior in the environment. The **Environment** includes the external conditions that allow the spread of the disease, such as geography, demography, weather, or social habits. All these elements merge in a **Time** span and we can see an emerging disease or a seasonal disease.

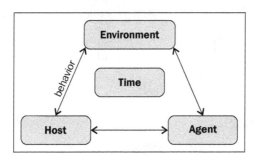

One of the most important concepts to take into account in epidemiology is the **basic reproduction ratio (R-0)**. It is a metric of the number of cases that one infected host can generate in its infectious period. When R-0 is less than 1, the infection will vanish in the long run. However, if R-0 is greater than 1, the infection will be able to spread in the host population. We can keep an endemic balance if R-0 in the susceptible population is equal to one.

The endemic, epidemic, and pandemic diseases can be explained as follows:

- **Endemic**: It is a disease that exists permanently in a particular population or geographic region
- **Epidemic**: It is a disease outbreak that infects many individuals in a population at the same time
- **Pandemic**: It occurs when an epidemic spreads at the worldwide level

For a complete reference to epidemiology concepts refer to *Introduction to Epidemiology 6th Edition* by *Ray M. Merrill*, Jones & Bartlett Learning (2012).

The epidemic models

When we want to describe how a pathogen or a disease is spread into a population, we need to create a model using mathematical, statistical, or computational tools. The most common model used in the epidemiology is **SIR (susceptible, infected, and recovered)** model, which was formulated in the paper *A Contribution to the Mathematical Theory of Epidemics* by *McKendrick and Kermack* published in 1927.

In the models presented in this chapter, we assume a closed population (without births or deaths) and that the demographics and socio-economic variables do not affect the spread of the disease.

The SIR model

The SIR epidemiological model describes the course of an infectious disease, as we can see in the following figure. Starting with a susceptible population (**S**), which comes into contact with an infected population (**I**), where the individual remains infected and once the infection period has passed, the individual is then in the recovered state (**R**):

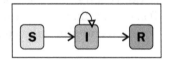

In this chapter we will use two different ways of solving the SIR model, a mathematical model with an **ordinary differential equations** (ODE) system and then with a computational model using a **cellular automaton** (CA). The two models should show a similar morphology (similar shape of the time series) in the three populations (susceptible, infected, and recovered) within the timeframe of the outbreak.

In the following figure we can see the ordinary differential equation system that represents the SIR model.

$$\text{(a)} \quad \frac{dS}{dt} = -\beta * S * I$$

$$\text{(b)} \quad \frac{dI}{dt} = \beta * S * I - \gamma * I$$

$$\text{(c)} \quad \frac{dR}{dt} = \gamma * I$$

Solving ordinary differential equation for the SIR model with SciPy

In order to observe the morphology of an infectious disease outbreak, we need to solve the SIR model. In this case, we will use the integrate method of the SciPy module to solve the ODE. In *Appendix, Setting Up the Infrastructure,* we can find the installation instructions for SciPy.

First, we need to import the required libraries scipy and pylab.

```
import scipy
import scipy.integrate
import pylab as plt
```

Then, we will define the SIR_model function, which will contain the ODE, with beta representing the transmission probability, gamma as the infected period, X[0] representing the susceptible population, and X[1] representing the infected population:

```
beta = 0.003
gamma = 0.1

def SIR_model(X, t=0):

  r = scipy.array([- beta*X[0]*X[1]
```

```
    ,    beta*X[0]*X[1]  -  gamma*X[1]
    ,    gamma*X[1] ])
   return r
```

Next, we will define the initial `parameters` (`[susceptible, infected, and recovered]`) and the `time` (number of days), then with the `scipy.integrate.odeint` function, we will solve the differential equations system:

```
if __name__ == "__main__":

  time = scipy.linspace(0, 60, num = 100)
  parameters = scipy.array([225, 1,0])
  X = scipy.integrate.odeint(SIR_model, parameters,time)
```

The result of `SIR_model` will look similar to the following list and will contain the status of the three populations (susceptible, infected, and recovered) for each step (days) during the outbreak:

```
[[  2.25000000e+02    1.00000000e+00    0.00000000e+00]
 [  2.24511177e+02    1.41632577e+00    7.24969630e-02]
 [  2.23821028e+02    2.00385053e+00    1.75121774e-01]
 [  2.22848937e+02    2.83085357e+00    3.20209039e-01]
 [  2.21484283e+02    3.99075767e+00    5.24959040e-01]
 . . .]
```

 All the codes and datasets of this chapter can be found in the author's GitHub repository at `https://github.com/hmcuesta/PDA_Book/tree/master/Chapter9`.

Finally, we will plot the three populations using `pylab`:

```
plt.plot(range(0, 100), X[:,0], 'o', color ="green")
plt.plot(range(0, 100), X[:,1], 'x', color ="red")
plt.plot(range(0, 100), X[:,2], '*', color ="blue")
plt.show()
```

In the following screenshot we can see the transition rates for the SIR model:

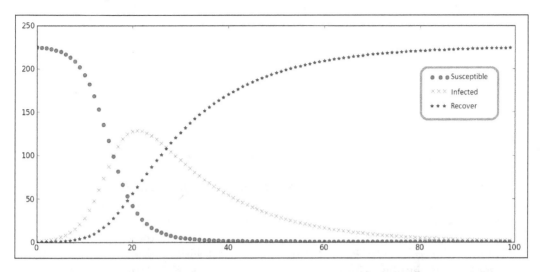

The SIRS model

The **SIRS (susceptible, infected, recovered, and susceptible)** model is an extension of the SIR model. In this case, the immunity acquired in the recovered status is eventually lost and the individual eventually comes back to the susceptible population. As we can see in the following screenshot, the SIRS model is cyclic. The SIRS model brings the opportunity to study other kinds of phenomenon a such as endemic diseases and seasonality effects. Some common examples of SIRS diseases are seasonal flu, measles, diphtheria, and chickenpox.

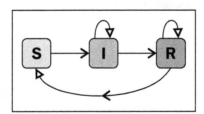

In the following figure we can see the ordinary differential equation system that represents the SIRS model:

$$\text{(a)} \quad \frac{dS}{dt} = -\beta * S * I + \sigma * R$$

$$\text{(b)} \quad \frac{dI}{dt} = \beta * S * I - \gamma * I$$

$$\text{(c)} \quad \frac{dR}{dt} = \gamma * I - \sigma * R$$

In order to solve the ODE (refer to the section *Solving ordinary differential equation for the SIR model with SciPy*), we need to create the SIRS_model function as shown in the following code, where the sigma variable represents the recovered period as shown in the ODE of the SIRS. We use the beta variable to represent the transmission probability, and gamma represents the infected period. Finally, we will use X[0] to represent the susceptible population, X[1] as the infected population, and X[2] as the recovered population.

```
beta = 0.003
gamma = 0.1
sigma = 0.1

def SIRS_model(X, t=0):

    r = scipy.array([- beta*X[0]*X[1] + sigma*X[2]
        ,  beta*X[0]*X[1] - gamma*X[1]
        ,  gamma*X[1] ] -sigma*X[2])
    return r
```

Modeling with cellular automata

Cellular automaton are mathematical and computational discrete models created by John von Neumann and Stanislaw Ulam. CA is represented as a grid where in each cell a small computation is performed. In CA we will share the process through all the small cells in the grid. CA shows behavior similar to biological reproduction and evolution. In this case, we can say that each cell is an individual in our population (grid) that will switch between states depending on its social interaction (contact rate). (Refer to SIR and SIRS models).

Seen as discrete simulations of dynamical systems, CA has been used for modeling in different areas such as traffic flow, encryption, growth of crystals, bird migration, and epidemic outbreaks. Stephen Wolfram, one of the most influential researchers in CA describes CA as follows:

> *"Cellular automata are sufficiently simple to allow detailed mathematical analysis, yet sufficiently complex to exhibit a wide variety of complicated phenomena."*

Cell, state, grid, and neighborhood

The basic element in a CA is the cell and it corresponds to a specific coordinate in a grid (or lattice). Each cell has a finite number of possible states and the current state will depend on a set of rules and the status of the surrounding cells (neighborhood). All the cells follow the same set of rules and when the rules are applied to the entire grid, we can say that a new generation is created.

The different kinds of neighborhoods are as follows:

- **Von Neumann**: It encompass the four cells orthogonally surrounding a central cell on a two-dimensional square grid

- **Moore**: It is the most common neighborhood and it encompasses the eight cells that surround the central cell in a two-dimensional grid

- **Moore Extended**: It has the same behavior as that of the Moore but in this case, we can extend the reach to different distances

- **Global**: In this case, the geometric distance is not considered and all the cells have the same probability to be reached by another cell. (Refer to the *Global stochastic contact model* section)

One of the most famous examples of CA is Conway's Game of Life. Where, in a two-dimensional lattice all the cells can be either dead or alive. In the following link we can see a D3.js visualization of the Game of Life http://bl.ocks.org/sylvaingi/2369589.

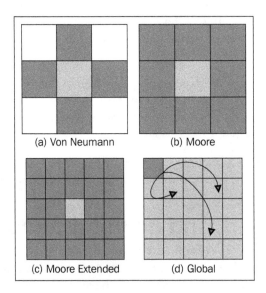

(a) Von Neumann   (b) Moore

(c) Moore Extended   (d) Global

Global stochastic contact model

For this model we will define the interaction between individuals in a homogeneous population. The contact is global and stochastic, this means that each cell has the same likelihood to be contacted by the other. In this model we do not consider the geographic distance, demographic, or the migratory pattern as constraints.

 We can find more information about the stochastic process at http://en.wikipedia.org/wiki/Stochastic_process.

Simulation of the SIRS model in CA with D3.js

In *Chapter 7*, *Predicting Gold Prices*, we already studied the basics of a random walk simulation. In this chapter, we will implement a CA in JavaScript using D3.js to simulate the SIRS model.

In the following screenshot we can see the interface of our simulator. It's a simple interface with a grid of 15 x 15 cells (225 total cells). An **Update** button that applies the rules to all the cells on the grid (step). A paragraph area that will show the status of different populations in the current step, for example, Susceptible: 35 Infected: 153 Recovered: 37 Step: 4. Finally, a **Statistics** button that writes a list with all the statistics of each step (susceptible, infected, recovered, and so on) into a text area for plotting purpose:

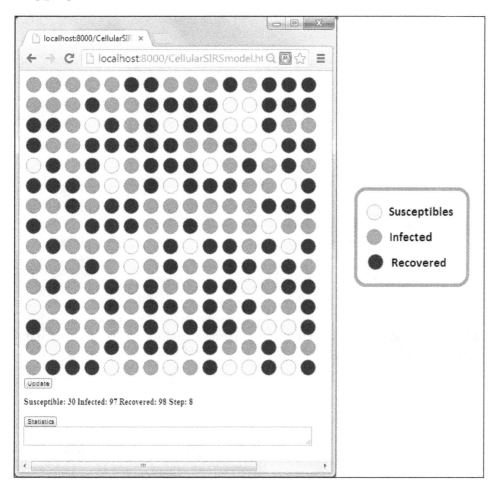

Inside the head tag we need to refer to the library:

```
<html>
<head>
  <script src="http://d3js.org/d3.v3.min.js"></script>
</head>
```

The code is mostly in JavaScript. First, we need to define the variables such as the grid, the list of colors, the number of rows and columns. Also, the SIRS model parameters such as the average number of contacts (avgContact), transmission probability (tProb), the initial number of infected (initialInfected), the infected period (timeInfection), and the recover period (timeRecover).

```
<body>
  <script type="text/javascript">

    var w = 600;
    var h = 600;

    var grid = [];
    var record = [];

    var colors = ["", "#F8F8F8", "#FF6633","000066"];
    var index = 0;
    var cols = 15;
    var rows =15;
    var nTimes = 0;
    var gridSize = (cols * rows);

    var avgContact = 4;
    var timeInfection = 2;
    var timeRecover = 4;
    var initialInfected = 10;
    var tProb = 0.2;
```

Now, we will fill the grid with susceptible cells using the push function. Each of the cells will contain an array with the coordinates, a unique index, the start status 1 (susceptible), and the period of time (in the initial state there is no period hence we assign 0).

```
for(var i=0; i <rows; i++ ){
  for(var j=0;j<cols;j++){
    grid.push([i*40,
      j*40,
      "circle-"+index++,
```

```
        1,
        0]);
    }
}
```

Then, we need to define the size of the new SVG width and height (600 x 600 pixels), which inserts a new `<svg>` element before the closing `</body>` tag:

```
var svg = d3.select("body")
    .append("svg")
    .attr("width", w)
    .attr("height", h)
    .append("g")
    .attr("transform","translate(20,20)");
```

Next, we need to generate the `circle` elements and add them to `svg`, then with the `data(grid)` function, for each value in data we will call the `.enter()` function and add a `circle` element. D3 allows selecting groups of elements for manipulation through the `selectAll` function. We will use each cell array inside the `grid` list to define coordinates (`cx`, `cy`), `color`, and `id`:

```
svg.selectAll("circle")
    .data(grid)
    .enter()
    .append("circle")
    .attr("id", function(d) {
        return d[2];
    })
    .attr("cx", function(d) {
        return d[0];
    })
    .attr("cy", function(d) {
        return d[1];
    })
    .attr("r", function(d) {
        return 15;
    })
    .attr("fill", colors[1])
    .attr("stroke", "#666");
```

Next, we will create the `init` function. It will be called only when we refresh the web page. The `init` function will randomly insert the initial number of infected cells in the CA:

```
function init(){
    for(var x = 0; x < initialInfected; x++ ){
```

```
            var i = Math.round(Math.random() * (gridSize-1));
            var cell = grid[i];
            if(cell[3]==1){
              cell[3] = 2;
              cell[4] = timeInfection;
            }
            grid[i] = cell;
        }
      prepareStep();
      }


    init();
```

In the `prepareStep` function we will refill all the circles with their new status color (`colors[cell[3]]`). We will use the function `svg.select` to select one element by its id (`cell[2]`) and apply the new style. The `prepareStep` function also counts the number of individuals in each of the three populations and shows them in the paragraph tag (`status`). Finally, the function stores the statistic of the current step into the record list:

```
    function prepareStep(){
    noSus = 0;
      noInfected = 0;
      noRecover = 0;
      nTimes++;

      for(var i = 0; i < gridSize; i++){
        var cell = grid[i];
        svg.select("#"+cell[2]).style("fill", colors[cell[3]]);

        if(cell[3] == 1){
          noSus++;
        }else if(cell[3] == 2){
          noInfected++;
        }else if(cell[3] == 3){
          noRecover++;
        }
      }
      record.push([noSus,noInfected,noRecover]);
        document.getElementById("status").innerHTML =
          " Suseptibles: "+ noSus+
          " Infected: "+noInfected+
          " Recovered: "+noRecover+
          " Times: "+ nTimes;
    }
```

The `nextStep` function will apply the rules defined in the SIRS model to each cell to define their new status. We will use the relative ID of the cells instead of their coordinates as it would be easier to reach the cell by its position in the list (0 to 224).

```
function nextStep(){

    for(var i=0; i < gridSize; i++){
```

We will take each `cell` one by one and apply their average number of contacts with the other cells:

```
var cell = grid[i];
```

We will check if the cell has a recovered status (3) and if the cell still has a time period in this status, then we just decrement the recovered period by one. However, if the recovered period is zero, we will perform the transition to the susceptible status (1):

```
if(cell[3]==3){

  if(cell[4] > 0){
    cell[4] = cell[4] - 1;
      }else{
    cell[3] = 1;
    cell[4] = 0;
  }

}else{
```

Now, if the status is 1 or 2 (susceptible or infected), we need to make random contacts and we compare the status of the first `cell` with the status of the second cell (`sCell`). If they have the same status, then we continue with the next contact. If either of the cells are infected, the other cell is exposed to the transmission probability (`tProb`) and if it's infected, then the cell is updated in the grid:

```
for(var j=0;j < avgContact ;j++){

  var sId = Math.round(Math.random() *
  (gridSize-1));
  var sCell = grid[sId];

  if(cell[3] == sCell[3]){
    continue;
```

```
}else if (cell[3] == 2 && sCell[3] == 1){

    if(Math.random() <= tProb){

        sCell[3] = 2;
        sCell[4] = timeInfection;

    }

}else if (cell[3] == 1 && sCell[3] == 2){
    if(Math.random() <= tProb){

        cell[3] = 2;
        cell[4] = timeInfection;

    }
}
    grid[sId] = sCell;
}
}
```

Next, if the cell is in the infected (2) status, then we will check if the period is over. In this case, we perform the transition to the recovered status. Otherwise, we just decrease the timer of the infectious period (cell[4]).

```
if(cell[3] == 2 && cell[4] == 0 ){

    cell[3] = 3;
    cell[4] = timeRecover;

}else if(cell[3] == 2 && cell[4] > 0 ){

    cell[4] = cell[4] - 1;

}

grid[i] = cell;
}
}
```

The `update` function triggers the new step for the CA by calling the `nextStep` function and the `prepareStep` function:

```
function update(){
   nextStep();
   prepareStep();
}
```

The `statistics` function writes the record list with the statistics of the simulation into the text area tag (`txArea`):

```
function statistics(){

   document.getElementById("txArea").value = ""+record;


}
</script>
```

Finally, we create the entire HTML code needed by the interface, the `Update` button, paragraph area (`status`), the `Statistics` button, and the text area (`txArea`).

```
<div id="option">
<input name="updateButton"
       type="button"
       value="Update"
       onclick="update()" />
</div>
<p id="status">Current Statistics</p>

<input name="updateButton"
       type="button"
       value="Statistics"
       onclick="statistics()" />
</br>
<textarea id=txArea
          cols = "70">
</textarea>
</body>
</html>
```

In the following screenshot we can observe the progression of the outbreak in the steps 1, 3, 6, 9, 11, and 14. We can appreciate how the SIRS model is applied to the grid:

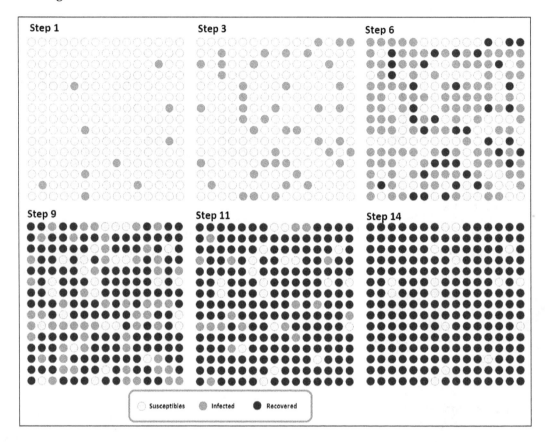

Now, we will copy the `record` list from the text area and we will visualize them in python with the small script shown as follows:

First, we will import the `pylab` and `numpy` modules:

```
import pylab as plt
import numpy as np
```

Then, we will create a `numpy` array with the record list:

```
data = np.array([215,10,. . .])
```

Next, in order to plot each population, we will reshape the array with `numpy` using the `reshape` method. The first parameter is `-1` because we don't know in advance how many steps are present and the second parameter defines the length of the array as 3 (susceptible, infected, and recovered).

```
result = data.reshape(-1,3)
```

The resultant array will look similar to the following array:

```
[[215  10    0]
 [153  72    0]
 [ 54 171    0]
 [  2 223    0]
 [  0 225    0]
 [  0 178   47]
 [  0  72  153]
 [  0   6  219]
 [  0   0  225]
 [ 47   0  178]
 [153   0   72]
 [219   0    6]
 [225   0    0]]
```

Finally, we use the `plot` method to display the visualization:

```
length = len(result)
plt.plot(range(0,length), result[:,0], marker = 'o', lw = 3,
color="green")
plt.plot(range(0,length), result[:,1], marker = 'x', linestyle = '--',
lw = 3, color="red")
plt.plot(range(0,length), result[:,2], marker = '*', linestyle =
':',lw = 3, color="blue")
plt.show()
```

In the following screenshot we can see the three populations throughout the time until all the cells come back to susceptible:

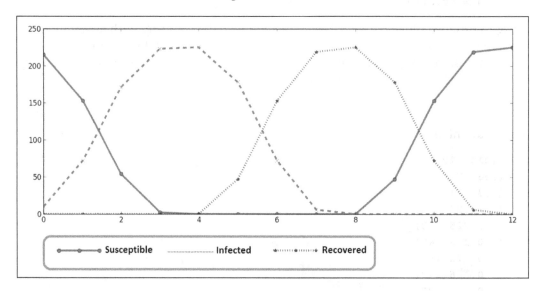

We can also play with the parameters such as the infectious period, the initial number of infected individuals, the transmission probability, or the recovered period. In the following screenshot we simulate the SIR model by increasing the recovered period into a large number and as we can observe, the result is highly similar to the result given by the mathematical model (ODE). Refer to the *Solving ordinary differential equation for the SIR model with SciPy* section:

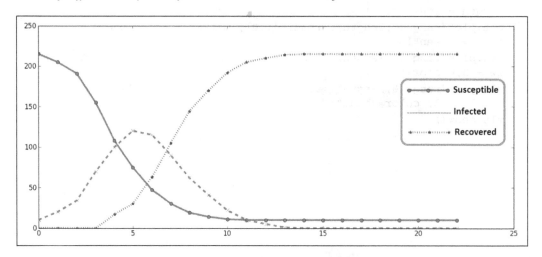

All the codes and datasets of this chapter can be found in the author's GitHub repository at https://github.com/ hmcuesta/PDA_Book/tree/master/Chapter9.

Summary

In this chapter, we introduced the basic concepts of epidemiology and two basic epidemic models (SIR and SIRS). Then, we learned how to model and solve an ordinary differential equations system for epidemic models. Finally, we developed a basic simulator implementing a cellular automaton of the SIRS model. We tried different parameters and got interesting results. Of course, these examples are only for educational purpose and if we need to model a real disease, we will need an epidemiologist to provide the accurate and real parameters.

In the next chapter, we will learn how to visualize and work with graphs from social networking sites.

10
Working with Social Graphs

In this chapter we introduce the most basic features of graph analytics. Initially, we distinguish the structure of a graph and a social graph, and how to obtain our friends' graph from Facebook. Then, we present some of the basic operations with graph—such as Degree and Centrality. Finally, we work in a graph representation using Gephi and we will create our own visualization in D3.js for our friends' graph.

In this chapter we will cover:

- Social Networks Analysis
- Acquiring my Friends list from Facebook
- Representing graphs with Gephi
- Statistical analysis of my graph (Degree and Centrality)
- Graph visualization with D3.js

Structure of a graph

A graph is a set of nodes (or vertices) and links (or edges). Each link is a pair of node references (such as source or target). Links may be considered as directed or undirected, depending if the relationship is mutual or not. The most common way to computationally represent a graph is by using an adjacency matrix. We use the index of the matrix as a node identifier and the value of the coordinates to represent whether there exists a link (the value is 1) or not (the value is 0). The links between nodes may have a scalar value (weight) to define a distance between the nodes. Graphs are widely used in Sociology, Epidemiology, Internet, Government, Commerce, and Social networks to find groups and information diffusion.

Graph analytics can be split into three categories:

- Structural algorithms
- Traversal algorithms
- Pattern-matching algorithms

Undirected graph

In the undirected graph, there is no distinction between the nodes source and target. As we can observe in the following figure the adjacency matrix is symmetric, which means that the relationship between nodes is mutual. This is the kind of graph used in Facebook, where we are friends with other nodes (symmetric relationship).

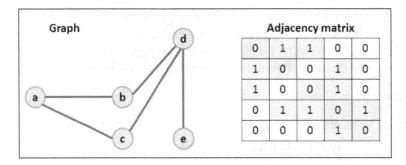

Directed graph

In the directed graph we find direction between the source node and the target node represented by an arrow, this creates an asymmetric (one-way) relationship. In this case we will have two different kinds of degree, In and Out. This can be observed in the adjacency matrix, which is not symmetric. This is particularly useful in networks such as Twitter, where we have followers, and not friends. It means that the relationship is not mutual by default and we will have two degrees, In (Followers) and Out (Following).

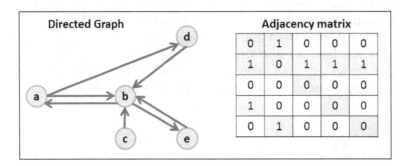

Social Networks Analysis

The **Social Networks Analysis (SNA)** is not a new technique; sociologists have been using it for a long time to study human relationships (sociometry), find communities, and to simulate how information or a disease is spread in a population.

With the rise of social networking sites such as Facebook, Twitter, LinkedIn, and so on, the acquisition of large amounts of social network data has become easier. We can use SNA to get an insight about customer behavior or unknown communities. It is important to say that this is not a trivial task and we will face problems with sparse data and a lot of noise (meaningless data). We need to understand, how to distinguish between false correlation and causation. A good start is by knowing our graph through visualization and statistical analysis.

The social networking sites bring us the opportunities to ask questions that otherwise are too hard to approach, because polling enough people is time-consuming and expensive.

In this chapter we will obtain our social network's graph from the Facebook (FB) website, in order to visualize the relationships between our friends. Then we will learn how to get an insight about the proportions of the nongraph data provided by FB such as gender or likes. Next, we will explore the distribution and centrality of our friends' relationships in our graph. Finally, we will create an interactive visualization of our graph using `D3.js`.

Acquiring my Facebook graph

In Facebook the friends represent nodes and the relationship between two friends represents links but we can get a lot more information from it, such as gender, age, post list, likes, political affiliation, religion, and so on. And Facebook provides us with a complete **Application Programming Interface (API)** to work with its data. You may visit the following link for more information:

`https://developers.facebook.com/`

Another interesting option is the Stanford Large Network Dataset Collection, where we can find social networks' datasets well formatted and anonymized for educational proposes. Visit the following link for more information:

`http://snap.stanford.edu/data/`

Using the anonymized data, it is possible to determine whether two users have the same affiliations, but not what their individual affiliations represent.

Using Netvizz

In this chapter we don't get into the use of the Facebook API. The easiest method to get our Friends list is by using a third-party application. Netvizz is a Facebook app developed by Bernhard Rieder, which allows exporting social graph data to GDF and tab formats. Netvizz may export information about our friends such as gender, age, locale, posts, and likes.

In order to get your social graph from Netvizz, you need to access the following link and give access to your Facebook profile:

```
https://apps.facebook.com/netvizz/
```

As shown in the following screenshot, we will create a GDF file from our personal friend network by clicking on the link **here** in **Step 2**.

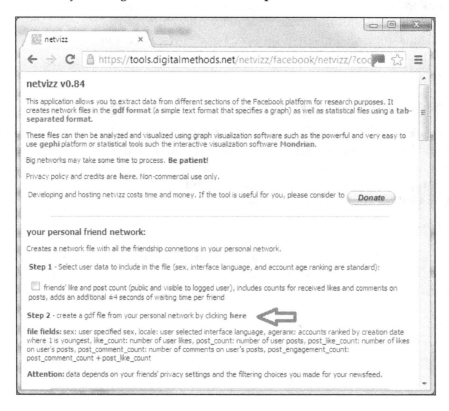

Then we will download the GDF (Graph Modeling Language) file which is a simple text format for a graph representation and is easy to use in Gephi (see the next section). Netvizz will give us the number of nodes and edges (links); finally we will click on the **gdf file** link, as we can see in the following screenshot:

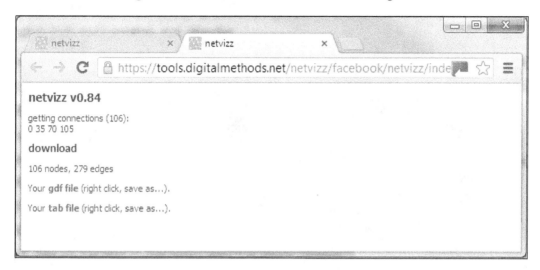

The output file `myFacebookNet.gdf` will look as follows:

```
nodedef>name VARCHAR,label VARCHAR,gender VARCHAR,locale VARCHAR,agerank
INT
23917067,Jorge,male,en_US,106
23931909,Haruna,female,en_US,105
35702006,Joseph,male,en_US,104
503839109,Damian,male,en_US,103
532735006,Isaac,male,es_LA,102
. . .
edgedef>node1 VARCHAR,node2 VARCHAR
23917067,35702006
23917067,629395837
23917067,747343482
23917067,755605075
23917067,1186286815
. . .
```

In the following figure we can see the visualization of the graph (106 nodes and 279 links). The nodes represent my friends and the links represent how my friends are connected between them. The graph is visualized with Gephi and the Force Atlas layout (refer to the section *Representing graphs with Gephi*).

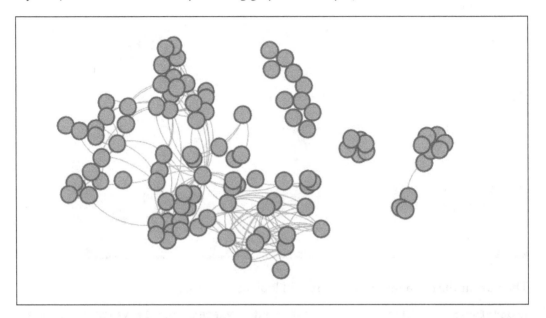

Netvizz can also obtain other kind of graphs such as **your like network:**, in this option Netvizz creates a graph of your friends and their likes (both friends and liked object(s) are nodes). If we scroll down the Netvizz interface we can find a section shown in the following screenshot. To create your like network, we just need to click on the **here** link:

your like network:

Create a bipartite network (gdf file) from your friends and their likes (both users and liked objects are nodes) **here**. Only liked pages are provided, not external objects. Count on waiting about a second per friend.

In this case Netvizz will create a graph with 106 friends, 6,388 different liked objects (6,494 nodes), and 7,965 links. Then to download the graph we just need to click on the **gdf file** link. As we can observe in the following figure the graph generated, in this case, is much denser than the friend graph. The graph is visualized with Gephi and the Force Atlas layout (refer to the section *Representing graphs with Gephi*).

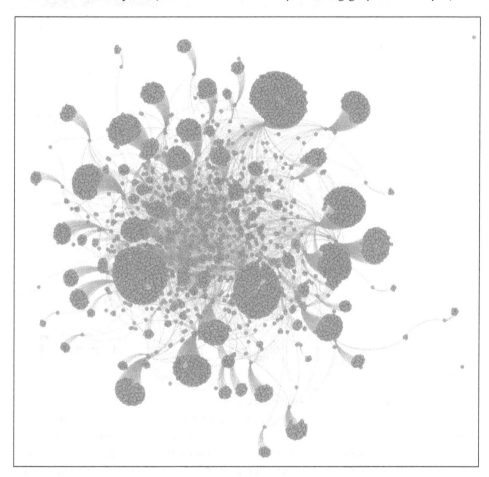

Representing graphs with Gephi

Gephi is an open source software for visualizing and analyzing large networks graphs which runs on Windows, Linux, and Mac OS X. We can freely download Gephi from its website listed as follows. For installation instructions please refer to the *Appendix, Setting up the Infrastructure*.

```
https://gephi.org/users/download/
```

To visualize your social network graph, you just need to open Gephi, click on the **File** menu and select **Open** then we just need to look up and select our file myFacebookNet.gdf and click on the **Open** button. Then, we can see our graph as shown in the following screenshot:

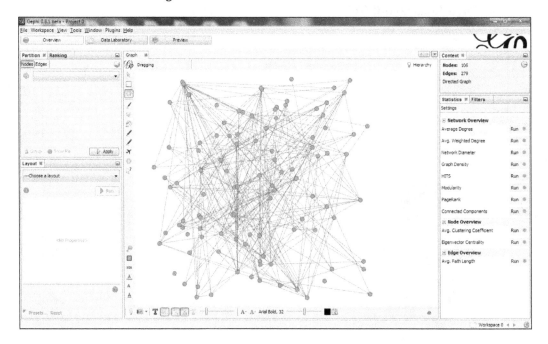

 For complete reference documentation about Gephi, please refer to the link https://gephi.org/users/.

In the interface we can see the **Context** tab, which shows us the number of **Nodes** and **Edges**. We can show Node labels by clicking on the **T** icon in the bottom of the window. Finally, we can apply different layout algorithms by selecting in the **---Choose a layout** dropdown in the **Layout** tab. Once the visualization is ready, we can click on the **Preview** button to get a better look at it and we can export the visualization to PDF, SVG, or PNG formats.

In the following screenshot we see the preview visualization of the graph using the Fruchterman-Reingold algorithm, which is a force-directed layout algorithm. The force-directed layouts are a family of algorithms for drawing graphs in dimensional spaces (2D or 3D), in order to represent the nodes and links of the graph in an aesthetical way.

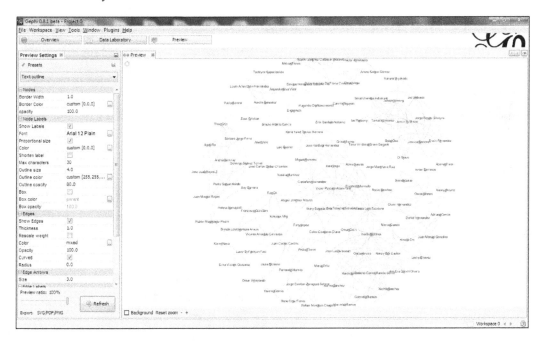

For more information about, Fruchterman-Reingold layout algorithm you can visit the link http://wiki.gephi.org/index.php/Fruchterman-Reingold.

Statistical analysis

We can easily find some information from our Facebook graph, such as the number of friends and individual data of each one. However, there are many questions that we can't get directly from the site, such as male to female ratio, how many of my friends are Republicans, or who is my best friend? These questions can be easily answered with a few lines of code and some basic statistical analysis. In this chapter we will start with male to female ratio, because we already have the gender value in the GDF file obtained from Netvizz.

For simplicity in the code examples, we will split the myFacebookNet.gdf file into two CSV files, one for the nodes (nodes.csv) and one for the links (links.csv).

Male to female ratio

In this example, we will use the gender value of the nodes.csv file and get the male to female ratio in a pie chart visualization.

The file nodes.csv will look as follows:

```
nodedef>name VARCHAR,label VARCHAR,gender VARCHAR,locale VARCHAR,agerank
INT
23917067,Jorge,male,en_US,106
23931909,Haruna,female,en_US,105
35702006,Joseph,male,en_US,104
503839109,Damian,male,en_US,103
532735006,Isaac,male,es_LA,102
```

. . .

1. Firstly, we need to import the required libraries. See *Appendix, Setting Up the Infrastructure,* for installation instructions of numpy and pylab.

   ```
   import numpy as np
   import operator
   from pylab import *
   ```

2. The numpy function, genfromtxt, will obtain only the gender column from the nodes.csv file, using the usecols attribute in the str format.

   ```
   nodes = np.genfromtxt("nodes.csv",
                   dtype=str,
                   delimiter=',',
                   skip_header=1,
                   usecols=(2))
   ```

3. Then we will use the function countOf from the operator module and ask for how many 'male' are in the list nodes.

   ```
   counter = operator.countOf(nodes, 'male')
   ```

4. Now, we just get the proportions between male and female in percentage.

   ```
   male = (counter *100) / len(nodes)
   female = 100 - male
   ```

5. Now, we make square `figure` and `axes`.

```
figure(1, figsize=(6,6))
ax = axes([0.1, 0.1, 0.8, 0.8])
```

6. Then, the slices will be ordered and plotted counter-clockwise.

```
labels = 'Male', 'Female'
ratio = [male,female]
explode=(0, 0.05)
```

7. Using the function `pie` we define the parameters of the chart such as explode, labels, and `title`.

```
pie(ratio,
    explode=explode,
    labels=labels,
    title('Male to Female Ratio',
        bbox={'facecolor':'0.8', 'pad':5})
```

8. Finally, with the function `show`, we execute the visualization.

```
show()
```

In the following figure we can see the pie chart. In this case we observe 54.7 percent male and 45.3 percent female:

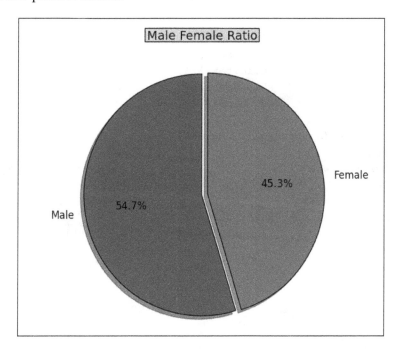

Degree distribution

The degree of a node is the number of connections (links) with other nodes. In the case of directed graphs, each node has two degrees: the out degree and the in degree. In the undirected graph, the relationship is mutual, so we just have a single degree for each node. In the following code snippet we get the source node and target node references from the file `links.csv`. Then we create a single list to merge the two lists (target and source). Finally, we get a dictionary (`dic`) of how many times each node appears in the list and we plot the result in a bar chart using `matplotlib`.

The file `links.csv` will look as follows:

```
edgedef>node1 VARCHAR,node2 VARCHAR
23917067,35702006
23917067,629395837
23917067,747343482
23917067,755605075
23917067,1186286815
    .    .    .
```

The complete code snippet looks as follows:

```
import numpy as np
import matplotlib.pyplot as plt

links = np.genfromtxt("links.csv",
                      dtype=str,
                      delimiter=',',
                      skip_header=1,
                      usecols=(0,1))
dic = {}
for n in sorted(np.reshape(links,558)):
    if n not in dic:
        dic[n] = 1
    else:
        dic[n] += 1
plt.bar(range(95),list(sort.values()))
plt.xticks(range(95), list(sort.keys()), rotation=90)
plt.show()
```

In the following figure, we can observe the degree of each node in the graph and there are 11 nodes that do not present any connection. In this example, from 106 total nodes in the graph we only consider the 95 nodes, which at least a have degree of one.

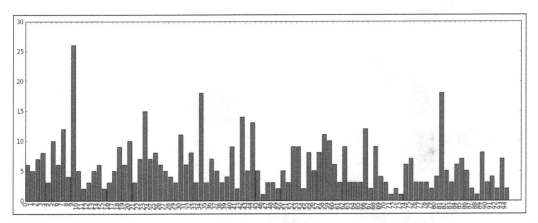

Histogram of a graph

Now, we will explore the structural task of the graph through its histogram. We will create a dictionary (`histogram`) that will contain, how many nodes have degree of one, two to 26 that is the maximum degree that can be reached by a node in this graph. Then, we will visualize the histogram using a scatter plot.

```
histogram = {}

for n in range(26):
    histogram[n] = operator.countOf(list(dic.values()), n)

plt.bar(list(histogram.keys())),list(histogram.values()))
plt.show()
```

In the following figure, we can see the histogram of our graph. The logical question here is "What does the histogram tell us about the graph?" The answer is that we can see a pattern in the histogram, and that many nodes in the graph have small degree and decreases. As we move along the X-axis, we can observe that most nodes have a degree of 3. In the pattern we can appreciate that it becomes less likely that a new node comes with a high degree, this is congruent with the Zipfian distribution. Most human-generated data presents this kind of distribution, such as words in vocabulary, letters in alphabet, and so on.

> For more information about, Zipfian distribution you can visit the link `http://en.wikipedia.org/wiki/Zipf's_law`.

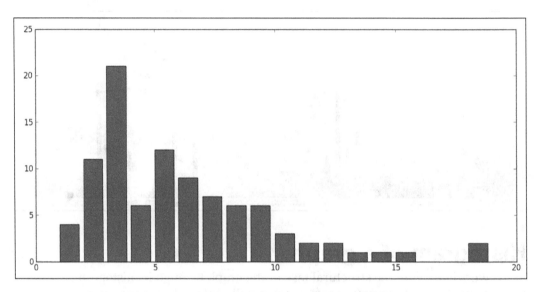

Other common pattern in graphs is exponential distribution and is frequently presented in random graphs.

Centrality

If we want to understand the importance of an individual node in the graph we need to define its centrality which is a relative measure of how important a node is within a graph. There are several ways to find centrality, such as closeness (average length of all its shortest paths) or betweenness (the fraction of all shortest paths that pass through a certain node). In this case, we will define centrality as the strongest connected node and we will prove this hypothesis through a direct data exploration.

In the following code snippet, we sort the dictionary by its value using a lambda function then we reverse the order to get the biggest degree, in the beginning.

```
sort = sorted(dic.items(), key=lambda x: x[1], reverse=True)
```

The result list, `sort`, will look as follows:

```
[('100001448673085', 26),
 ('100001452692990', 18),
 ('100001324112124', 18),
```

```
('100002339024698', 15),
('100000902412307', 14),
 .   .   . ]
```

In the following screenshot, we can see the graph visualized in Gephi with Yifan Hu Layout algorithm. With a direct data exploration we can color the node with the apparent highest degree and we can say that it is the central node. Now, perform the following steps:

1. In Gephi interface click on the **Ranking** tab.

2. Select the **Degree** option.

3. In the combobox, pick a **Color** and select the highest **Range** (26/27).

4. Click on the **Apply** button (refer the underlined options in the following screenshot).

We can also color the first degree contacts of the central node and see that it is strongly connected between groups. We can do this by selecting the Painter tool in Gephi and clicking on all the nodes related with the central node. In fact, we find that is the same node with the highest degree obtained by the sorted process (ID 100001448673085).

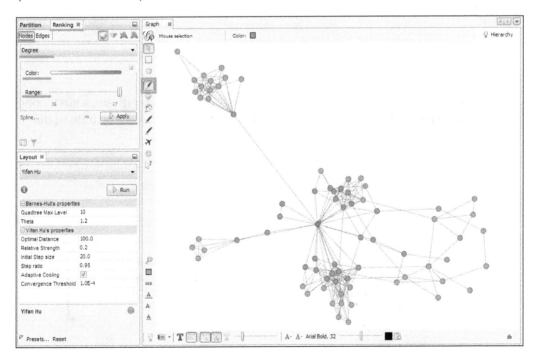

We can create our own centrality algorithm, based not just in the degree (number of connections of the node). For example, we can find centrality in the number of shares and likes of certain nodes' posts. This means that a node with a lower degree may have a bigger impact in the information diffusion process or a particular node is strongly connected in between different groups; that's the beauty of social networks.

Transforming GDF to JSON

Gephi is an excellent tool to get easy and fast results. However, if we want to present the graph interactively on a website, we need to implement a different kind of visualization. In order to work with the graph in the web, we need to transform our GDF file to JSON format.

1. Firstly, we need to import the libraries numpy and json. For more information about JSON format, refer to *Chapter 2, Working with Data*.

```
import numpy as np
import json
```

2. The numpy function, genfromtxt, will obtain only the ID and name from the nodes.csv file using the usecols attribute in the 'object' format.

```
nodes = np.genfromtxt("nodes.csv",
                      dtype='object',
                      delimiter=',',
                      skip_header=1,
                      usecols=(0,1))
```

3. Then, the numpy function, genfromtxt, will obtain links with the source node and target node from the links.csv file using the usecols attribute in the 'object' format.

```
links = np.genfromtxt("links.csv",
                      dtype='object',
                      delimiter=',',
                      skip_header=1,
                      usecols=(0,1))
```

The JSON format used in the D3.js Force Layout graph implemented in this chapter requires transforming the ID (for example, 100001448673085) into a numerical position in the list of nodes.

4. Then, we need to look for each appearance of the ID in the links and replace them by their position in the list of nodes.

```
for n in range(len(nodes)):
    for ls in range(len(links)):
        if nodes[n][0] == links[ls][0]:
            links[ls][0] = n

        if nodes[n][0] == links[ls][1]:
            links[ls][1] = n
```

5. Now, we need to create a dictionary "data" to store the JSON file.

```
data ={}
```

6. Next, we need to create a list of nodes with the names of the friends in the format as follows:

    `"nodes": [{"name": "X"},{"name": "Y"},. . .]` and add it to the data dictionary.

```
lst = []
for x in nodes:
    d = {}
    d["name"] = str(x[1]).replace("b'","").replace("'","")
    lst.append(d)

data["nodes"] = lst
```

7. Now, we need to create a list of links with the source and target in the format as follows:

    `"links": [{"source": 0, "target": 2},{"source": 1, "target": 2},. . .]` and add it to the data dictionary.

```
lnks = []

for ls in links:
    d = {}
    d["source"] = ls[0]
    d["target"] = ls[1]
    lnks.append(d)

data["links"] = lnks
```

8. Finally, we need to create the file, newJson.json, and write the data dictionary in the file with the function dumps of the json library.

```
with open("newJson.json","w") as f:
    f.write(json.dumps(data))
```

Neo4j is a robust (fully ACID) transactional property graph database. For more information you can visit the link about Neo4j at http://www.neo4j.org/.

The file newJson.json will look as follows:

```
{"nodes": [{"name": "Jorge"},
        {"name": "Haruna"},
        {"name": "Joseph"},
        {"name": "Damian"},
        {"name": "Isaac"},
        . . .],
 "links": [{"source": 0, "target": 2},
        {"source": 0, "target": 12},
        {"source": 0, "target": 20},
        {"source": 0, "target": 23},
        {"source": 0, "target": 31},
        . . .]}
```

Graph visualization with D3.js

D3.js provides us with the d3.layout.force() function that use the Force Atlas layout algorithm and help us to visualize our graph. Refer to *Chapter 3, Data Visualization*, for instructions on how to create D3.js visualizations.

1. Firstly, we need to define the CSS style for the nodes, links, and node labels.

```
<style>

.link {
  fill: none;
  stroke: #666;
  stroke-width: 1.5px;
}

.node circle
{
  fill: steelblue;
  stroke: #fff;
  stroke-width: 1.5px;
```

```
}

.node text
{
  pointer-events: none;
  font: 10px sans-serif;
}
</style>
```

2. Then, we need to refer the d3js library.

```
<script src="http://d3js.org/d3.v3.min.js"></script>
```

3. Then, we need to define the width and height parameters for the svg container and include into the body tag.

```
var width = 1100,
    height = 800

var svg = d3.select("body").append("svg")
    .attr("width", width)
    .attr("height", height);
```

4. Now, we define the properties of the force layout such as gravity, distance, and size.

```
var force = d3.layout.force()
    .gravity(.05)
    .distance(150)
    .charge(-100)
    .size([width, height]);
```

5. Then, we need to acquire the data of the graph using the JSON format. We will configure the parameters for nodes and links.

```
d3.json("newJson.json", function(error, json) {
  force
      .nodes(json.nodes)
      .links(json.links)
      .start();
```

 For a complete reference about the d3js Force Layout implementation, visit the link https://github.com/mbostock/d3/wiki/Force-Layout.

6. Then, we define the links as lines from the `json` data.

```
var link = svg.selectAll(".link")
      .data(json.links)
    .enter().append("line")
      .attr("class", "link");

var node = svg.selectAll(".node")
      .data(json.nodes)
    .enter().append("g")
      .attr("class", "node")
      .call(force.drag);
```

7. Now, we define the node as circles of size 6 and include the labels of each node.

```
node.append("circle")
    .attr("r", 6);

node.append("text")
      .attr("dx", 12)
      .attr("dy", ".35em")
      .text(function(d) { return d.name });
```

8. Finally, with the function, `tick`, run step-by-step the force layout simulation.

```
force.on("tick", function()
{
    link.attr("x1", function(d) { return d.source.x; })
        .attr("y1", function(d) { return d.source.y; })
        .attr("x2", function(d) { return d.target.x; })
        .attr("y2", function(d) { return d.target.y; });

    node.attr("transform", function(d)
    {
            return "translate(" + d.x + "," + d.y + ")";
    })
    });
});
</script>
```

In the following screenshot we can see the result of the visualization. In order to run the visualization we just need to open a command terminal and run the following command:

```
>>python -m http.server 8000
```

After that we just need to open a web browser and type the direction
`http://localhost:8000/ForceGraph.html`. In the HTML page, we can see our
Facebook graph with a gravity effect and we can interactively drag-and-drop the nodes.

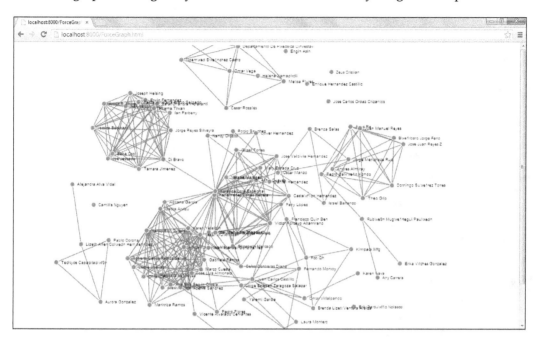

 All the code and datasets of this chapter may be found in the
author's GitHub repository at the link `https://github.`
`com/hmcuesta/PDA_Book/Chapter10`.

The complete code of the visualization is listed as follows:

```
<meta charset="utf-8">
<style>

.link
{
  fill: none;
  stroke: #666;
  stroke-width: 1.5px;
}
.node circle
{
  fill: steelblue;
  stroke: #fff;
  stroke-width: 1.5px;
```

```
    }

    .node text
    {
      pointer-events: none;
      font: 10px sans-serif;
    }
    </style>
    <body>
    <script src="http://d3js.org/d3.v3.min.js"></script>
    <script>

    var width = 1100,
        height = 800

    var svg = d3.select("body").append("svg")
        .attr("width", width)
        .attr("height", height);

    var force = d3.layout.force()
        .gravity(.05)
        .distance(150)
        .charge(-100)
        .size([width, height]);

    d3.json("newJson.json", function(error, json) {
      force
          .nodes(json.nodes)
          .links(json.links)
          .start();

      var link = svg.selectAll(".link")
          .data(json.links)
        .enter().append("line")
          .attr("class", "link");

      var node = svg.selectAll(".node")
          .data(json.nodes)
        .enter().append("g")
          .attr("class", "node")
          .call(force.drag);

      node.append("circle")
```

```
        .attr("r", 6);

    node.append("text")
        .attr("dx", 12)
        .attr("dy", ".35em")
        .text(function(d) { return d.name });

    force.on("tick", function()
  {
      link.attr("x1", function(d) { return d.source.x; })
          .attr("y1", function(d) { return d.source.y; })
          .attr("x2", function(d) { return d.target.x; })
          .attr("y2", function(d) { return d.target.y; });

      node.attr("transform", function(d)
    { return "translate(" + d.x + "," + d.y + ")"; });
    });
  });
  </script>
  </body>
```

Summary

In this chapter we worked on how to obtain and visualize our Facebook graph, applying some layouts with Gephi such as Force Atlas and Fruchterman-Reingold. Then we introduced some of the statistical methods to get aggregate information such as degree, centrality, distribution, and ratio. Finally, we developed our own visualization tool with D3.js, transforming the data from GDF into JSON.

In the next chapter, we will present a short introduction to the Twitter API to retrieve, visualize, and analyze tweets. Then, we will proceed to perform a sentiment analysis.

11
Sentiment Analysis of Twitter Data

In this chapter we will see how to perform sentiment analysis over Twitter data. Initially, we introduce the **Twitter API** with Python. Then, we distinguish the basic elements of a sentiment classification. Finally, we present the **Natural Language Toolkit (NLTK)** to implement the tweets' sentiment analyzer.

In this chapter we will cover:

- The anatomy of Twitter data
- Using OAuth to access Twitter API
- Getting started with Twython:
    - Simple search/query
    - Working with timelines
    - Working with followers
    - Working with places and trends

- Sentiment classification:
    - Effective norms for English words
    - Text corpus

- Get started with Natural Language Toolkit (NLTK)
    - Bag of words
    - Naïve Bayes
    - Sentiment analysis of tweets

In *Chapter 4, Text Classification*, we presented a basic introduction to text classification. In this chapter, we will perform a sentiment analysis of tweets to rate the emotional value (positive or negative) using classification with Naive Bayes method.

Sentiment analysis can be used to find patterns in the opinion of the population such as where people are happier or what is the public perception about a brand new product.

With the proliferation of social networking websites, we can see what people are talking about in real-time and on a large scale. However, we need to be cautious because the social networks tend to be noisy, that is why in this case, we will need as much data as we can get in order to obtain a true representation of what people think.

The anatomy of Twitter data

Twitter is a social networking website, which provides a micro-blogging service for sharing text messages up to 140 characters (or tweets). We can retrieve a variety of data from Twitter such as **tweets, followers, favorites, direct messages**, and **trending topics**.

We can create a new Twitter account using the following link:

```
http://twitter.com
```

Tweet

Tweet is the name of the 140-character long text message. However, we can get more information than the text message itself such as **date and time, links, user mentions (@), hash tags (#), retweets count, locale language, favorites count**, and **geocode**. In the following screenshot, we can observe a tweet retweeted 1001 times, marked as favorite 336 times with a hashtag (**#NBAFinals**) and user mentions (**@Spurs** and **@MiamiHEAT**):

Followers

The users on Twitter can follow other users creating a directed graph (See *Chapter 10, Working with Social Graphs*) with a lot of possibilities for analysis such as centrality and community clustering. In this case, the relationship is not mutual by default, so on Twitter we have two kinds of degrees; in and out. This can be very useful when we want to find the most influent individual in a group or which individual is in between different groups.

> We can follow the Twitter's engineering team's blog at `https://engineering.twitter.com/`.

Trending topics

Twitter trends are words or hashtags with a high popularity among Twitter users at a specific moment and/or place. Trending topics is a big area for data analysis such as how to detect trends and predict future trends. These are main topics in information diffusion theory. In the following screenshot, we can see the dialog box used to change tailored trends (trends based on your location and who you follow on Twitter) simply by changing your location:

Using OAuth to access Twitter API

In order to have access to the Twitter API, we will use a **token-based authentication** system. Twitter applications are required to use OAuth, which is an open standard for authorization. OAuth allow the Twitter users to enter their username and password in order to obtain four strings (token). The token allows the users to connect with the Twitter API without using their username and password. In this chapter, we will use the current version of **Twitter REST API 1.1**, released on June 11, 2013, which established the use of OAuth authentication as mandatory, for retrieving data from Twitter.

> For more information about token-based authentication systems, please refer to `http://bit.ly/bgbmnK`.

First, we need to visit `https://dev.twitter.com/apps` and sign in with our Twitter username and password as is shown in the following screenshot:

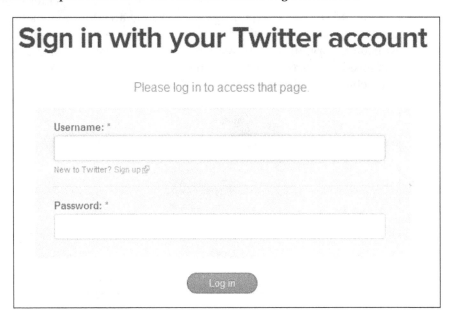

Then, we click on the **Create a new application** button (see the following screenshot) and enter the application details:

- **Name**: `PracticalDataAnalysisBook` (can be anything you like; however, you cannot use the word Twitter in the name)

- **Description**: `Practical Data Analysis Book Examples` (can be anything you like)

- **Website**: Can be your personal blog or website
- **Callback URL**: Can be left blank

Next, we need to enter the CAPCHA and click on the **Create** button.

My applications

Create a new application

Looks like you haven't created any applications yet!

Finally, on the next details screen, we will click on **Create my access token** (sometimes you need to manually refresh the page after a few seconds). In the following screenshot, we can see our four strings for authentication; **Consumer key**, **Consumer secret**, **Access token**, and **Access token secret**.

OAuth settings

Your application's OAuth settings. Keep the "Consumer secret" a secret. This key should never be human-readable in your application.

Access level	Read-only
	About the application permission model
Consumer key	
Consumer secret	
Request token URL	https://api.twitter.com/oauth/request_token
Authorize URL	https://api.twitter.com/oauth/authorize
Access token URL	https://api.twitter.com/oauth/access_token
Callback URL	None
Sign in with Twitter	No

Your access token

Use the access token string as your "oauth_token" and the access token secret as your "oauth_token_secret" to sign requests with your own Twitter account. Do not share your oauth_token_secret with anyone.

Access token	
Access token secret	
Access level	Read-only

Now, we may use this access token with multiple user timelines on multiple websites using the Twitter Search API. However, this is restricted to 180 requests/queries per 15 minutes.

 We can find more information about the Twitter Search API, limitations, best practices, and rate limits at `https://dev.twitter.com/docs/using-search`.

Getting started with Twython

In this chapter, we will use Twython 3, which is a Python wrapper of the Twitter API 1.1. We can download the latest version of `twython` from `pypi` Python website, `https://pypi.python.org/pypi/twython`.

Then, we need to unzip and open the `twython` folder. Finally we install the `twython` module using the following command:

```
>>> python3 setup.py install
```

Or, we can also install Twython through `easy_install` using the following command.

```
>>> easy_install twython
```

We can find complete reference documentation of Twython at `https://twython.readthedocs.org/en/latest/index.html`.

You can find a complete list of Twitter libraries for several programming languages such as Java, C#, Python, and so on at `https://dev.twitter.com/docs/twitter-libraries`.

Simple search

In this example, we will perform a search of the word python and we will print the complete list of statuses in order to understand the format of the retrieved tweets.

First, we need to import the Twython object from the `twython` library:

from twython import Twython

Then, we need to define the four strings created using OAuth (see section *Using OAuth to Access Twitter API*):

```
ConsumerKey   = "..."
ConsumerSecret = "..."
AccessToken = "..."
AccessTokenSecret = "..."
```

Now, we need to instantiate the `Twython` object giving the access token string as the parameters:

```
twitter = Twython(ConsumerKey,
    ConsumerSecret,
    AccessToken,
    AccessTokenSecret)
```

Next, we will perform the search, using the `search` method, specifying the search query text in the keyword argument `q`:

```
result = twitter.search(q="python")
```

 Twython converts the JSON sent to us from Twitter to a naïve python object. However, if the authentication fails, the search will retrieve an error message as follows:

```
{"errors":[{"message":"Bad Authentication data",
    "code":215}]}
```

Finally, we will iterate in the `result["statuses"]` list and print each status (tweet):

```
for status in result["statuses"]:
    print(status)
```

The output of each status is retrieved in a JSON-like structure and will look as follows:

```
{'contributors': None,
 'truncated': False,
 'text': 'La théorie du gender.... Genre Monty python ! http://t.
co/3nTUhVR9Xm',
 'in_reply_to_status_id': None,
 'id': 355755364802764801,
 'favorite_count': 0,
 'source': '<a href="http://twitter.com/download/iphone"
rel="nofollow">Twitter for iPhone</a>',
 'retweeted': False,
```

```
'coordinates': None,
'entities': {'symbols': [],
        'user_mentions': [],
        'hashtags': [],
        'urls': [{'url': 'http://t.co/3nTUhVR9Xm',
            'indices': [46, 68],
            'expanded_url': 'http://m.youtube.com/watch?feature=youtube_
gdata_player&v=ePCSA_N5QY0&desktop_uri=%2Fwatch%3Fv%3DePCSA_
N5QY0%26feature%3Dyoutube_gdata_player',
            'display_url': 'm.youtube.com/watch?feature=…'}]},
            'in_reply_to_screen_name': None,
        'in_reply_to_user_id': None,
        'retweet_count': 0,
        'id_str': '355755364802764801',
        'favorited': False,
        'user': {'follow_request_sent': False,
            'profile_use_background_image': True,
            'default_profile_image': False,
            'id': 1139268894,
            'verified': False,
            'profile_text_color': '333333',
            'profile_image_url_https': 'https://si0.twimg.com/profile_
images/3777617741/d839f0d515c0997d8d18f55693a4522c_normal.jpeg',
            'profile_sidebar_fill_color': 'DDEEF6',
            'entities': {'url': {'urls': [{'url':
'http://t.co/7ChRUG0D2Y',

        'indices': [0, 22],

        'expanded_url': 'http://www.manifpourtouslorraine.fr',

        'display_url': 'manifpourtouslorraine.fr'}]},

        'description': {'urls': []}},
'followers_count': 512,
'profile_sidebar_border_color': 'C0DEED',
'id_str': '1139268894',
```

```
'profile_background_color': 'C0DEED',
'listed_count': 9,
'profile_background_image_url_https':
'https://si0.twimg.com/images/themes/theme1/bg.png',
'utc_offset': None,
'statuses_count': 249,
'description': "ON NE LACHERA JAMAIS, RESISTANCE !!\r\nTous
nés d'un homme et d'une femme\r\nRetrait de la loi Taubira
!\r\nRestons mobilisés !",
'friends_count': 152,
'location': 'moselle',
'profile_link_color': '0084B4',
'profile_image_url':
'http://a0.twimg.com/profile_images/3777617741/d839f0d515c099
  7d8d18f55693a4522c_normal.jpeg',
'following': False,
'geo_enabled': False,
'profile_banner_url':
'https://pbs.twimg.com/profile_banners/1139268894/1361219172',
'profile_background_image_url':
'http://a0.twimg.com/images/themes/theme1/bg.png',
'screen_name': 'manifpourtous57',
'lang': 'fr',
'profile_background_tile': False,
'favourites_count': 3,
'name': 'ManifPourTous57',
'notifications': False,
'url': 'http://t.co/7ChRUG0D2Y',
'created_at': 'Fri Feb 01 10:20:23 +0000 2013',
'contributors_enabled': False,
'time_zone': None,
'protected': False,
'default_profile': True,
'is_translator': False},
'geo': None,
'in_reply_to_user_id_str': None,
```

```
'possibly_sensitive': False,
'lang': 'fr',
'created_at': 'Fri Jul 12 18:27:43 +0000 2013',
'in_reply_to_status_id_str': None,
'place': None,
'metadata': {'iso_language_code': 'fr', 'result_type':
'recent'}}
```

We can also restrict the result by navigating through the structure of the JSON result. For example, to get only `user` and `text` of the status, we can modify the `print` command as follows:

```
for status in result["statuses"]:
    print("user: {0} text: {1}".format(status["user"]["name"],
    status["text"]))
```

The output of the first five statuses will look as follows:

user: RaspberryPi-Spy text: RT @RasPiTV: RPi.GPIO Basics Part 2, day 2 -
Rev checking (Python & Shell) http://t.co/We8PyOirqV

user: Ryle Ploegs text: I really want the whole world to watch Monty
Python and the Holy Grail at least once. It's so freaking funny.

user: Matt Stewart text: Casual Friday night at work... #snakes #scared
#python http://t.co/WVld2tVV8X

user: Flannery O'Brien text: Kahn the Albino Burmese Python enjoying the
beautiful weather :) http://t.co/qvp6zXrG60

user: Cian Clarke text: Estonia E-Voting Source Code Made Public
http://t.co/5wCulH4sht - open source, kind of! Python & C http://t.
co/bo3CtukYoU

. . .

Navigating through the JSON structure helps us to get only the information that we need for our applications. We may pass multiple keyword arguments and also specify the result type with the `result_type="popular"` parameter.

> We can find a complete reference of GET search/tweets at https://dev.
> twitter.com/docs/api/1.1/get/search/tweets.

Working with timelines

In this example, we will show how to retrieve our own timeline and a different user's timeline.

First, we need to import and instantiate the `Twython` object from the `twython` library:

```
from twython import Twython
ConsumerKey       = "..."
ConsumerSecret    = "..."
AccessToken       = "..."
AccessTokenSecret = "..."
twitter = Twython(ConsumerKey,
  ConsumerSecret,
  AccessToken,
  AccessTokenSecret)
```

Now, to get our own timeline we will use the `get_home_timeline` method.

```
timeline = twitter.get_home_timeline()
```

Finally, we will iterate the timeline and print the `user name`, `created at`, and `text`.

```
for tweet in timeline:
    print(" User: {0} \n Created: {1} \n Text: {2} "
      .format(tweet["user"]["name"],
        tweet["created_at"],
        tweet["text"]))
```

The first five results of the code will look as follows:

User: Ashley Mayer

Created: Fri Jul 12 19:42:46 +0000 2013

Text: Is it too late to become an astronaut?

User: Yves Mulkers

Created: Fri Jul 12 19:42:11 +0000 2013

Text: The State of Pharma Market Intelligence http://t.co/v0f1DH7KZB

User: Olivier Grisel

Created: Fri Jul 12 19:41:53 +0000 2013

Text: RT @stanfordnlp: Deep Learning Inside: Stanford parser quality improved with new CVG model. Try the englishRNN.ser.gz model. http://t.co/jE...

user: Stanford Engineering

Created: Fri Jul 12 19:41:49 +0000 2013

Text: Ralph Merkle (U.C. Berkeley), Martin Hellman (#Stanford Electrical #Engineering) and Whitfield Diffie… http://t.co/4y7Gluxu8E

User: Emily C Griffiths

Created: Fri Jul 12 19:40:45 +0000 2013

Text: What role for equipoise in global health? Interesting Lancet blog: http://t.co/2FA6ICfyZX

. . .

On the other hand, if we want to retrieve a specific user's timeline such as `stanfordeng`, we will use the `get_user_timeline` method with the `screen_name` parameter to define the user selected, and we can also restrict the number of results to five, using the `count` parameter:

```
tl = twitter.get_user_timeline(screen_name = "stanfordeng",
  count = 5)
for tweet in tl:
  print(" User: {0} \n Created: {1} \n Text: {2} "
    .format(tweet["user"]["name"],
      tweet["created_at"],
      tweet["text"]))
```

The first five statuses of Stanford Engineering (@stanfordeng) timeline will look as follows:

Created: Fri Jul 12 19:41:49 +0000 2013

Text: Ralph Merkle (U.C. Berkeley), Martin Hellman (#Stanford Electrical #Engineering) and Whitfield Diffie… http://t.co/4y7Gluxu8E

User: Stanford Engineering

Created: Fri Jul 12 15:49:25 +0000 2013

Text: @nitrogram W00t!! ;-)

User: Stanford Engineering

Created: Fri Jul 12 15:13:00 +0000 2013

Text: Stanford team (@SUSolarCar) to send newest creation, solar car #luminos for race in Australia: http://t.co/H5bTSEZcYS. via @ paloaltoweekly

User: Stanford Engineering

Created: Fri Jul 12 02:50:00 +0000 2013

Text: Congrats! MT @coursera: Coursera closes w 43M in Series B. Doubling in size to focus on mobile, apps platform & more! http://t.co/ WTqZ7lbBhd

User: Stanford Engineering

```
Created: Fri Jul 12 00:57:00 +0000 2013
Text: Engineers can really benefit from people who can make intuitive or
creative leaps. ~Stanford Electrical Engineering Prof. My Le  #quote
```

 We can find the complete reference of the home_timeline and user_timeline methods at the following links:
- http://bit.ly/nEpIW9
- http://bit.ly/QpgvRQ

Working with followers

In this example, we will show how to retrieve the list of followers of specific Twitter users.

First, we need to import and instantiate the Twython object from the twython library:

```
from twython import Twython
ConsumerKey       = "..."
ConsumerSecret    = "..."
AccessToken       = "..."
AccessTokenSecret = "..."
twitter = Twython(ConsumerKey,
   ConsumerSecret,
   AccessToken,
   AccessTokenSecret)
```

Next, we will return the list of followers with the get_followers_list method using screen_name (username) or user_id (Twitter user ID):

```
followers = twitter.get_followers_list(screen_name="hmcuesta")
```

Next, we iterate over the followers["users"] list and print all the followers:

```
for follower in followers["users"]:
   print(" {0} \n ".format(follower))
```

Each user will look as follows:

```
{'follow_request_sent': False,
 'profile_use_background_image': True,
 'default_profile_image': False,
 'id': 67729744,
 'verified': False,
 'profile_text_color': '333333',
```

'profile_image_url_https':
'https://si0.twimg.com/profile_images/374723524/iconD_normal.gif',
'profile_sidebar_fill_color': 'DDEEF6',
'entities': {'description': {'urls': []}},
'followers_count': 7,
'profile_sidebar_border_color': 'C0DEED',
'id_str': '67729744',
'profile_background_color': 'C0DEED',
'listed_count': 0,
'profile_background_image_url_https':
'https://si0.twimg.com/images/themes/theme1/bg.png',
'utc_offset': -21600,
'statuses_count': 140,
'description': '',
'friends_count': 12,
'location': '',
'profile_link_color': '0084B4',
'profile_image_url':
'http://a0.twimg.com/profile_images/374723524/iconD_normal.gif',
'following': False,
'geo_enabled': False,
'profile_background_image_url':
'http://a0.twimg.com/images/themes/theme1/bg.png',
'screen_name': 'jacobcastelao',
'lang': 'en',
'profile_background_tile': False,
'favourites_count': 1,
'name': 'Jacob Castelao',
'notifications': False,
'url': None,
'created_at': 'Fri Aug 21 21:53:01 +0000 2009',
'contributors_enabled': False,
'time_zone': 'Central Time (US & Canada)',
'protected': True,
'default_profile': True,
'is_translator': False}

Finally, we will print only the user (`screen_name`), name, and the number of tweets (`statuses_count`).

```
for follower in followers["users"]:
    print(" user: {0} \n name: {1} \n Number of tweets: {2} \n"
        .format(follower["screen_name"],
            follower["name"],
            follower["statuses_count"]))
```

The first five followers result of the preceding code will look as follows:

```
user: katychuang
name: Kat Chuang, PhD
number of tweets: 1991

user: fractalLabs
name: Fractal Labs
number of tweets: 105

user: roger_yau
name: roger yau
number of tweets: 70

user: DataWL
name: Data Without Limits
number of tweets: 1168

user: abhi9u
name: Abhinav Upadhyay
number of tweets: 5407
```

 We can find the complete reference of the `get_followers_list` method at `https://dev.twitter.com/docs/api/1.1/get/followers/list`.

Working with places and trends

In this example, we will retrieve the trending topics closest to a specific location. In order to specify the location, Twitter API uses the **WOEID (Yahoo! Where On Earth ID)**.

First, we need to import and instantiate the `Twython` object from the `twython` library:

```
from twython import Twython
ConsumerKey       = "..."
ConsumerSecret    = "..."
AccessToken       = "..."
AccessTokenSecret = "..."
twitter = Twython(ConsumerKey,
   ConsumerSecret,
   AccessToken,
   AccessTokenSecret)
```

Next, we will use `get_place_trends` and we define the place with the `id = (WOEID)` parameter:

```
result = twitter.get_place_trends(id = 23424977)
```

 We can find the complete reference of the `get_place_trends` method at `https://dev.twitter.com/docs/api/1.1/get/trends/closest`.

The easiest way to get the WOEID is through the console of **Yahoo! Query Language (YQL)**, which uses a SQL-like syntax; so if we want to find the WOEID of Denton Texas, the string query will look as follows:

```
select * from geo.places where text="Denton, TX"
```

We can find the console at the following link and we can test the string query by clicking on the **Test** button:

```
http://developer.yahoo.com/yql/console/
```

In the following screenshot, we can see the result of the query in a JSON format and pointed the `woeid` attribute with an arrow:

Finally, we will iterate the `result` list and print `name` of each `trend`:

```
if result:
    for trend in result[0].get("trends", []):
        print("{0} \n".format(trend["name"]))
```

The trending topics in Denton, TX will look as follows:

#20FactsAboutMyBrother

#ImTeamTwist

`Ho Lee Fuk`

#FamousTamponQuotes

#TopTenHoeQuotes

#BaeLiterature

`Cosart`

`KTVU`

`NTSB`

`Pacific Rim`

Sentiment classification

In sentiment classification, one message can be classified as either positive or negative. This is excellent to get insight about how the public think about a person, products, or services.

In this chapter, we will classify the tweets to get a personal positive or negative feeling. It is important to clarify that tweets are limited to 140-characters length with a very casual language and in many cases the message may be very noisy with usernames, links, repeated letters, and emoticons. However, Twitter provides a way to get feedback about large amount of topics in real-time. We can see sample tweets as follows:

```
"Photoshop, I hate it when you crash " - Negative
"@Ms_HipHop im glad ur doing weeeell " - Positive
```

The general process of the sentiment classification is presented in the following screenshot. We start extracting the features (words) from the training data (Text Corpus). Then, we need to train the classifier with a bag of words, which is a list of words and its frequency in the text. For example, the word great appears 32 times in the positive texts (tweets). Next, we will perform a query using the Twitter API, and extract the features of the resulted statuses to classify them as either positive or negative.

In the following sections, we will describe each part of the sentiment classification process and will use the Naive Bayes classifier implemented in the NLTK (Natural Language Toolkit) library, in order to classify if a tweet is either positive or negative.

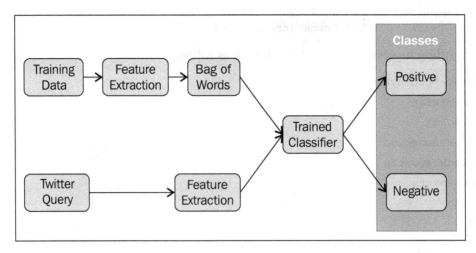

Affective Norms for English Words

Affective Norms for English Words (**ANEW**) has been developed for the Center of the Study of Emotion and Attention of the University of Florida. The ANEW provides a set of normative emotional ratings as a text corpus for a large number of words in the English language. These sets of verbal materials have been rated in terms of pleasure, arousal, and dominance, in order to create a standard for use in studies of emotion and attention. Even if the terms of use of ANEW are for nonprofit education proposes only and cannot be used for commercial purpose, it is an interesting option. We can get more information about ANEW from `http://csea.phhp.ufl.edu/media/anewmessage.html`.

Text corpus

Text corpus is a set of processed and labeled texts of a single or multiple languages, used in linguistics for statistical analysis. A corpus may contain text messages or paragraphs and we will split these into either **unigrams** (words individually) or **bigrams** (two words attached with only one meaning). We will use only English unigrams in this chapter.

When we want to create a corpus, we will get as much data as we can. Sometimes performing a feature reduction may help to increase the accuracy by avoiding usernames, links, repeated letters emoticons (for example, :D) from the text. The website **Sentiment140** provides us with a human-labeled corpus with over 1,600,000 tweets; with three polarities labeled with a 0 for negative, 2 for neutral, and 4 for positive. The corpus also provides the tweet id, date, user that tweeted, and the text. We can download the corpus from `http://help.sentiment140.com/for-students`.

We will use only polarities 0 and 4 (negative and positive) and the Sentiment140 corpus cvs file will look as follows:

```
"0","1824518676","Sun May 17 01:43:54 PDT 2009","NO_
QUERY","marielmilo","my tummy hurts "

"0","1824519186","Sun May 17 01:44:02 PDT 2009","NO_
QUERY","jtetsuya","Watching Himitsu no Hanazono again... I wish stories
like this never had endings  Amuro Namie's - Baby Don't Cry http://
tinyurl.com/qr4ros"

"0","1824519390","Sun May 17 01:44:06 PDT 2009","NO_
QUERY","jackspencer","I hate getting up late... and to find that I have a
day of solid german revision to do "

"0","1824519463","Sun May 17 01:44:07 PDT 2009","NO_
QUERY","SummerSlacking","I missed my friend's birthday party. I feel bad
and douchey. I haven't seen her in over a year. "
```

```
"0","1824519867","Sun May 17 01:44:14 PDT 2009","NO_
QUERY","Beggeesgirl","just got home in bed, but no phonecall from begee.
uh oh, im kinda worried! "

. . .

"4","2193577828","Tue Jun 16 08:38:52 PDT 2009","NO_
QUERY","ogreenthumb","@crgrs359 Skip the aquarium and check out these
fish   A lot cheaper lol http://bit.ly/21QbBv"

"4","2193577852","Tue Jun 16 08:38:52 PDT 2009","NO_QUERY","dacyj","@
GroleauNET Yeah I'm being an ass today "

"4","2193577870","Tue Jun 16 08:38:52 PDT 2009","NO_
QUERY","stephmartinez","@OHTristaN it's sunoudy "

"4","2193577904","Tue Jun 16 08:38:53 PDT 2009","NO_
QUERY","heartcures","@kbonded Newsflash: It worked "
```

Getting started with Natural Language Toolkit (NLTK)

NLTK is a powerful Python library for computational linguistics and text classification. NLTK include about 50 corpora and lexical resources such as Wordnet. NLTK is the most used tool for natural language processing in Python. It includes powerful algorithms for text tokenization, parsing, semantic reasoning, and text classification. We can find a complete guide of NLTK from `http://nltk.org/`.

To install NLTK, we just need to download the executable file from the website for windows and use `easy_install` in Linux distributions.

 We may need to install PyYaml in order to use NLTK. We can download PyYaml from `http://pyyaml.org/wiki/PyYAML`.

NLTK defines four basic classifiers:

- Naive Bayes
- Maximum entropy (or Logistic regression)
- Decision tree
- Conditional exponential

 In this chapter, we will use NLTK 3.0, which supports Python 3. However, it's still in alpha release (Sept 2013) and is likely to contain bugs. We can download the NLTK 3 from `http://nltk.org/nltk3-alpha/`.

Bag of words

Bag of words model is used to turn a document into an unsorted list of words, commonly used for classifying texts by getting the frequency of a word in a document. We will use the frequency as a feature for the training of the classifier. In NLTK, we have methods such as `nltk.word_tokenize` and `nltk.FreqDist` that make easier it to get the words and their frequency in a text.

In the following code, we can see how to import NLTK and the use of the `nltk.word_tokenize` method:

```
>>> import nltk
>>> nltk.word_tokenize("Busy day ahead of me. Also just remembered
that I left peah slices in the fridge at work on Friday. ")
```

```
['Busy', 'day', 'ahead', 'of', 'me.', 'Also', 'just', 'remembered',
'that', 'I', 'left', 'peah', 'slices', 'in', 'the', 'fridge', 'at',
'work', 'on', 'Friday', '.']
```

Naive Bayes

Naive Bayes is a simple model which works well to perform text classification. In the *Chapter 4, Text Classification*, we introduced the basic concepts of Naive Bayes algorithm. NLTK includes an implementation of Naive Bayes algorithm. In the following code, we will implement a Naive Bayes algorithm and we will use it to classify the tweets from a simple query using Twitter API.

First, we need to import the `nltk` library:

```
import nltk
```

Now, we will define three functions to get the bag of words (`bagOfWords`) and extract the frequency of each word in the tweets (`wordFeatures` and `getFeatures`):

```
def bagOfWords(tweets):
  wordsList = []
  for (words, sentiment) in tweets:
    wordsList.extend(words)
  return wordsList

def wordFeatures(wordList):
  wordList = nltk.FreqDist(wordList)
  wordFeatures = wordList.keys()
  return wordFeatures

def getFeatures(doc):
```

```
docWords = set(doc)
feat = {}
for word in wordFeatures:
  feat['contains(%s)' % word] = (word in docWords)
return feat
```

Next, we will define the corpus with two lists of positive and negative tweets. We will use 200 positive and 200 negative tweets extracted from the Sentiment140 human-labeled corpus. If we want to improve the accuracy of the algorithm, we can use a bigger corpus similar to the one that we mentioned in the section *Text Corpus*. However, in order to work with big amounts of data we should see *Chapter 12, Data Processing and Aggregation with MongoDB*.

```
positiveTweets = [('...', 'positive'),
   ('...', 'positive'), . . . ]



negativeTweets = [('. . .', 'negative'),
   ('...', 'negative'), . . .]
```

Then, we will create the corpus, merge the positive and the negative tweets, and extract the list of words using the `nltk.word_tokenize` method just excluding the words with less than three characters:

```
corpusOfTweets = []
for (words, sentiment) in positiveTweets + negativeTweets:
  wordsFiltered = [e.lower() for e in nltk.word_tokenize(words) if
len(e) >= 3]
  tweets.append((wordsFiltered, sentiment))
```

 NLTK already include several corpora, toy grammars, and trained models for deferent contexts such as movie reviews or people names. We can install NLTK data from `http://nltk.org/data.html`.

Now, we will get the features of all words:

```
wordFeatures = wordFeatures(bagOfWords(corpusOfTweets))
```

Next, we will get the training set using the `nltk.classify.apply_features` method:

```
training = nltk.classify.apply_features(getFeatures,
   corpusOfTweets)
```

Finally, we will train the Naïve Bayes algorithm as shown in the following code:

```
classifier = nltk.NaiveBayesClassifier.train(training)
```

We can get the most informative features of our classifier using the `show_most_informative_features` method. We can see the result in the following screenshot. This list shows the most frequent or informative words used by this classifier:

```
print(classifier.show_most_informative_features(32))
```

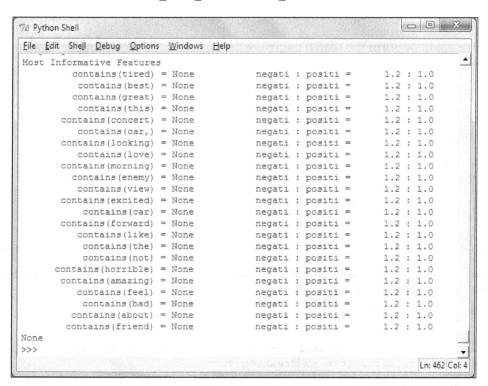

Sentiment analysis of tweets

Now we will perform a twitter search of the word Python and we will classify each tweet as positive or negative using the `classifier.classify` method (see section *Getting started with Twython*).

First, we need to import the `Twython` object from the `twython` library, and define the four strings created using OAuth:

```
from twython import Twython
ConsumerKey   = "..."
ConsumerSecret = "..."
AccessToken = "..."
AccessTokenSecret = "..."
```

Now, we need to instantiate the `Twython` object giving the access token string as the parameter:

```
twitter = Twython(ConsumerKey,
    ConsumerSecret,
    AccessToken,
    AccessTokenSecret)
```

Next, we will perform the search using the `search` method, specifying the search query text in the keyword argument `q`:

```
result = twitter.search(q="python")
```

Finally, we will iterate over the `result["statuses"]` list and will use the method to get the sentiment of each tweet:

```
for status in result["statuses"]:
  print("Tweet: {0} \n Sentiment: {1}"
    .format(status["text"],
    classifier.classify(extract_features
      (status["text"].split()))))
```

The output of the first five tweets and its sentiment classification will look as follows:

```
Tweet: RT @RasPiTV: RPi.GPIO Basics Part 2, day 2 - Rev checking (Python
&amp; Shell) http://t.co/We8PyOirqV

Sentiment: positive


Tweet: I really want the whole world to watch Monty Python and the Holy
Grail at least once. It's so freaking funny.

Sentiment: positive


Tweet: Casual Friday night at work... #snakes #scared #python http://t.
co/WVld2tVV8X

Sentiment: positive


Tweet: Kahn the Albino Burmese Python enjoying the beautiful weather :)
http://t.co/qvp6zXrG60

Sentiment: positive


Tweet: Estonia E-Voting Source Code Made Public http://t.co/5wCulH4sht -
open source, kind of! Python &amp; C http://t.co/bo3CtukYoU

Sentiment: positive
```

. . .

In this case, all the results from the search were classified as positive. Although we will need more tests and a larger training set, we can use the `classifier. accuracy` method to see the quality of our classifier. For this example, the accuracy is 73 percent, which is good for a short corpus.

Summary

In this chapter, we covered the basic functions of Twitter API from signing in with OAuth to location trends and how to perform simple queries. Then, we introduced the concepts of sentiment classification and developed a basic sentiment-analysis tool for tweets. There are a lot of ways to improve the classifier such as getting a bigger corpus and performing more complex queries. However, the accuracy of the example is good for educational propose.

In the next chapter, we will present the basic concepts of MongoDB and how we can perform aggregation queries with large amount of data.

12
Data Processing and Aggregation with MongoDB

Aggregation queries are a very common way to get summarized data by counting or adding features to our dataset. MongoDB provides us with different ways to get the aggregated data quickly and easily. In this chapter, we will explore the basic features of MongoDB as well as two ways to get summarized data using the group function and the aggregation framework.

In this chapter we will cover:

- Getting started with MongoDB:
  - Database
  - Collections
  - Documents
  - Mongo shell
  - Insert/Update/Delete operations
  - Queries

- Data Processing:
  - Data transformation with OpenRefine
  - Inserting documents with PyMongo

- Group
- The aggregation framework:
  - Pipeline
  - Expressions

In *Chapter 2, Working with Data*, we introduced the NoSQL (Not Only SQL) databases and their types (document-based, graph-based, and key-value stores). The NoSQL databases provide key advantages to the user such as scalability, high availability, and processing speed. Due to the distributed nature of the NoSQL technology, if we want to scale a NoSQL database we just need to add machines to the cluster to meet demand (horizontal scaling). Most of the NoSQL databases are open-source (such as MongoDB); which means that we can download, implement, and scale them for a very low cost.

Getting started with MongoDB

MongoDB is a very popular document-oriented NoSQL database. MongoDB provides a high-performance engine for storage and query retrieval. In a document-oriented database, we store the data into collections of documents, in this case JSON-like documents called **BSON (Binary JSON)**, which provide us with a dynamic-schema data structure. MongoDB implements functionalities such as **ad hoc queries**, **replication**, **load balancing**, **aggregation**, and **Map-Reduce**. MongoDB is perfect for an operational database. However, its capabilities as a transactional datasource are limited. We can see the similarities between the structures of a relational database (RDBMS) and MongoDB in the following diagram. You can find more information about MongoDB from its official website, `http://www.mongodb.org/`.

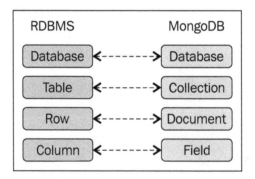

For complete reference about SQL databases, please refer to `http://www.w3schools.com/sql/sql_quickref.asp/`.

From the preceding figure, we can see that the internal structure of MongoDB is very similar to a relational database. However, in this case, we have a set of collections with BSON documents in it, without any previous schema defined and not all documents in a collection must have the same schema. For complete MongoDB installation instructions, we can see the *Installing and running MongoDB* section of *Appendix*, *Setting Up the infrastructure*.

Database

In MongoDB, a database is a physical container for our collections. Each database will create a set of files on the file system. In MongoDB, a database is created automatically on the fly when we save a document into a collection for first time. However, administration tools such **UMongo** allow us to create databases as is shown in the following screenshot:

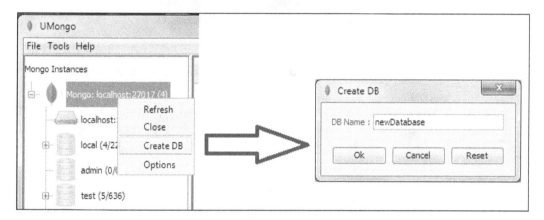

 You may find the complete UMongo installation instructions in the *Appendix*, *Setting Up the Infrastructure*, and we can find more information at http://edgytech.com/umongo/.

Additionally, we can see the available databases with the show dbs command and the result can be seen in the following screenshot:

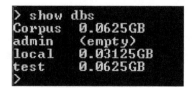

Collection

A collection is a group of documents. MongoDB will create the collection implicitly as is done with the database. As MongoDB uses a schema-less model, we must specify the database and the collection where it will be stored. MongoDB provides a JavaScript function `db.createCollection()` to create a collection manually and we may also create collections from the UMongo interface.

We can specify the database with the `use <database name>` command in the Mongo shell and we can see the available collections in the database with the `show collections` command as shown in the following screenshot:

```
> use Corpus
switched to db Corpus
> show collections
system.indexes
tweets
>
```

Collections can be split to distribute the collections documents across the MongoDB instances (shards). This process is called Sharding and allows a horizontal scaling.

> You may find considerations about data modeling in MongoDB at
> `http://docs.mongodb.org/manual/core/data-modeling/`.

Document

A document is a record in MongoDB and implements a schema-less model. This means that the documents are not enforced to have the same set of fields or structure. However, in practice the documents share a basic structure in order to perform queries and complex searches.

MongoDB use a document format similar to **JSON (JavaScript Object Notation)**, stored in a binary representation called **BSON**. For Python programmers, we will use the same dictionary structure to represent the JSON format as seen in *Chapter 2, Working with Data*. We may find the complete BSON specification at `http://bsonspec.org/`.

MongoDB uses a dot notation (`.`) to navigate through the JSON structure, to access a field in the document or subdocument. For example, `<subdocument>.<field>`.

Mongo shell

Mongo shell is an interactive JavaScript console for MongoDB. Mongo shell comes as a standard feature in the MongoDB. We also have the option to try a small version of mongo shell from the official website (see the following figure), good enough to start with MongoDB.

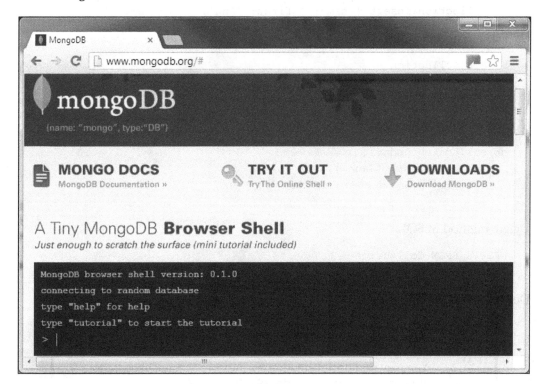

 You can find the FAQs about mongo shell at
http://docs.mongodb.org/manual/faq/mongo/.

Insert/Update/Delete

Now, we will explore the basic operations with MongoDB and will compare them with the analogous instructions in SQL just as references. If you already have some experience with relational database SQL language, this is going to be a natural transition.

Insert method in SQL:

```
INSERT INTO Collection (First_Name, Last_Name)
   Values ('Jan', 'Smith');
```

Insert method in MongoDB:

```
db.collection.insert({ name: { first: 'Jan', last: 'Smith' } )
```

Update method in SQL:

```
UPDATE Collection
SET First_Name = 'Joan'
WHERE First_Name = 'Jan';
```

Update method in MongoDB:

```
db.collection.update(
   { 'name.first': 'Jan' },
   { $set: { 'name.first': 'Joan' } }
)
```

Delete method in SQL:

```
DELETE FROM Collection
WHERE First_Name = 'Jan';
```

Delete method in MongoDB:

```
db.collection.remove( { 'name.first' : 'Jan' }, safe=True )
```

> You may find documentation for the Core MongoDB operations at http://docs.mongodb.org/manual/crud/.

Queries

In MongoDB, we can perform searches and retrieve data with two methods; `find` and `findOne`, both are listed as follows:

Selecting all elements from the Collection table in SQL:

```
SELECT * FROM Collection
```

Selecting all elements from the collection in MongoDB:

```
db.collection.find()
```

In the following screenshot, we can see the result of the find method in the mongo shell:

```
> db.test.data.find()
{ "_id" : ObjectId("51eedee2d341516bbfdbc6ff"), "name" : { "first" : "Jan", "las
t" : "Smith" } }
{ "_id" : ObjectId("51eedf0cd341516bbfdbc700"), "name" : { "first" : "Damian", "
last" : "Cuesta" } }
{ "_id" : ObjectId("51eedf17d341516bbfdbc701"), "name" : { "first" : "Isaac", "l
ast" : "Cuesta" } }
> _
```

Getting the number of documents retrieved by a query with SQL:

```
SELECT count(*) FROM Collection
```

Getting the number of documents retrieved by a query with MongoDB:

```
db.collection.find().count()
```

Query with a specific criteria with SQL:

```
SELECT * FROM Collection
WHERE Last_Name = "Cuesta"
```

Query with a specific criteria with MongoDB:

```
db.collection.find({"name.last":"Cuesta"})
```

In the following screenshot, we can see the result of the find method using specific criteria in the mongo shell:

```
> db.test.data.find({"name.last":"Cuesta"})
{ "_id" : ObjectId("51eedf0cd341516bbfdbc700"), "name" : { "first" : "Damian", "
last" : "Cuesta" } }
{ "_id" : ObjectId("51eedf17d341516bbfdbc701"), "name" : { "first" : "Isaac", "l
ast" : "Cuesta" } }
>
```

The findOne method retrieves a single document from the collection and does not return a list of documents (cursor). In the following screenshot, we can see the result of the findOne method in the mongo shell:

```
> db.test.data.findOne()
{
        "_id" : ObjectId("51eedee2d341516bbfdbc6ff"),
        "name" : {
                "first" : "Jan",
                "last" : "Smith"
        }
}
```

You can find documentation for read operations at
http://docs.mongodb.org/manual/core/read-operations/.

When we want to test the query operation and the timing of the query, we will use the `explain` method. In the following screenshot, we can see the result of the `explain` method to find the efficiency of the queries and index used.

In the following code we can see the use of the `explain` method in the `find` method:

```
db.collection.find({"name.last":"Cuesta"}).explain()
```

```
> db.test.data.find({"name.last":"Cuesta"}).explain()
{
        "cursor" : "BasicCursor",
        "isMultiKey" : false,
        "n" : 2,
        "nscannedObjects" : 3,
        "nscanned" : 3,
        "nscannedObjectsAllPlans" : 3,
        "nscannedAllPlans" : 3,
        "scanAndOrder" : false,
        "indexOnly" : false,
        "nYields" : 0,
        "nChunkSkips" : 0,
        "millis" : 0,
        "indexBounds" : {

        },
        "server" : "Hadoop-PC:27017"
}
```

Data preparation

In *Chapter 11, Sentiment Analysis of Twitter Data*, we explored how to create a bag of words from the `Tweets Sentiment140` dataset. In this chapter, we will complement the example by using MongoDB. First we will prepare and transform the dataset from CSV to a JSON format in order to add it into a MongoDB collection.

 We can download the Sentiment140 training and test data from
http://help.sentiment140.com/for-students.

We will download and open the test data, the columns represent sentiment, id, date, via, user, and text. The first five records will look like this:

```
4,1,Mon May 11 03:21:41 UTC 2009,kindle2,yamarama,@mikefish  Fair enough.
But i have the Kindle2 and I think it's perfect  :)

4,2,Mon May 11 03:26:10 UTC 2009, jquery,dcostalis,Jquery is my new best
friend.

4,3,Mon May 11 03:27:15 UTC 2009,twitter,PJ_King,Loves twitter

4,4,Mon May 11 03:29:20 UTC 2009,obama,mandanicole,how can you not love
Obama? he makes jokes about himself.

4,5,Mon May 11 05:22:12 UTC 2009,lebron,peterlikewhat,lebron and zydrunas
are such an awesome duo
```

The first problem that we can see is that the text field includes the comma (,) character in it. This will be a problem if we want to read the file from Python. In order to solve this problem we will perform a data preparation in OpenRefine before we start working with the file. See *Chapter 2, Working with Data*, for an introduction to OpenRefine.

Data transformation with OpenRefine

First we need to run OpenRefine (see *Appendix, Setting Up the Infrastructure*, for installation instructions) and import the `testdata manual 2009 06 14.csv` file. Then we will select the number of columns (separated by commas) and click on the **Create the project** button. In the following screenshot, we can see the interface of OpenRefine with six columns and we can rename the columns by clicking on each column, and then navigating to **Edit column | Rename this column**:

In order to delete the comma character from the text field, we need to click on the **test** column, and then navigate to **Edit Cells | Transform...**. Now, in the **Custom text transform on column text** window, we will use the `replace` function to eliminate all the commas from text, as is shown in the following screenshot.

In the following command, we can see the `replace` function from the **OpenRefine Expression Language** (**GREL**):

```
value.replace(",", "")
```

GREL implements a large selection of functions for strings, arrays, math, dates, and boolean. We can find more information at `https://github.com/OpenRefine/OpenRefine/wiki/GREL-Functions`:

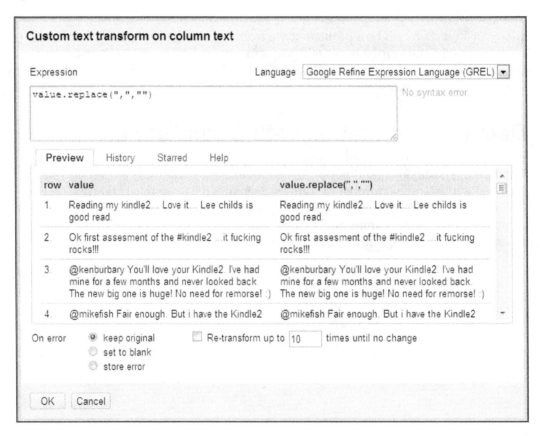

Finally to export the dataset into a JSON format, we will select the **Export** select box and then select **Templating**. Then we can see the **Templating Export** window (see following screenshot) where we can define the final structure and the row template in JSON format. Finally, we need to click on the **Export** button to download the test.json file into a file system location:

Templating Export

Prefix
```
{
  "rows" : [
```

Row Template
```
{
  "sentiment" : {{jsonize(cells["sentiment"].valu
  "id" : {{jsonize(cells["id"].value)}},
  "date" : {{jsonize(cells["date"].value)}},
  "via" : {{jsonize(cells["via"].value)}},
  "user" : {{jsonize(cells["user"].value)}},
  "text" : {{jsonize(cells["text"].value)}}
}
```

Row Separator
```
,
```

Suffix
```
  ]
}
```

```
{
  "rows" : [
    {
      "sentiment" : 4,
      "id" : 4,
      "date" : "Mon May 11 03:18:03 UTC 2009",
      "via" : "kindle2",
      "user" : "vcu451",
      "text" : "Reading my kindle2...  Love it... L
    },
    {
      "sentiment" : 4,
      "id" : 5,
      "date" : "Mon May 11 03:18:54 UTC 2009",
      "via" : "kindle2",
      "user" : "chadfu",
      "text" : "Ok first assesment of the #kindle2
    },
    {
      "sentiment" : 4,
      "id" : 7,
      "date" : "Mon May 11 03:21:41 UTC 2009",
      "via" : "kindle2",
      "user" : "yamarama",
```

Reset Template   Export   Cancel

Inserting documents with PyMongo

With the dataset in JSON format it will be much easier to insert the records into the MongoDB collection. In this chapter, we will use UMongo as a GUI (Graphic User Interface) tool, the Python module pymongo (see the *Appendix*, *Setting Up the Infrastructure*, for installation instructions of UMongo and PyMongo), and the json module:

```
import json
from pymongo import MongoClient
```

 You can find complete documentation about PyMongo at http://api.mongodb.org/python/current/.

We will establish connection in UMongo and we will create a new database by clicking on **localhost** and selecting **Create DB**, as shown in the following screenshot:

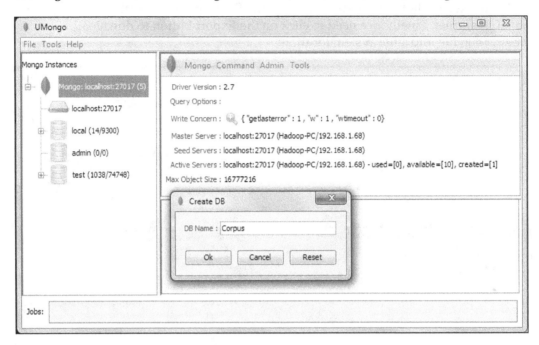

Next, we need to establish connection with `pymongo` using the connection function:

```
con = MongoClient()
```

Then, we will select the `Corpus` database:

```
db = con.Corpus
```

Now, we will select the `tweets` collection where all the documents will be stored:

```
tweets = db.tweets
```

Finally, we will open the `test.txt` file as structure of dictionaries with the `json.load` function:

```
with open("test.txt") as f:
    data = json.loads(f.read())
```

Then we will iterate all the rows and insert into the `tweets` collection:

```
for tweet in data["rows"]:
    tweets.insert(tweet)
```

The result can be seen in UMongo by navigating through the `Corpus` database and the `tweets` collection. Then we use the left-click and select the **find** option to retrieve all the documents in the collection as shown in the following screenshot:

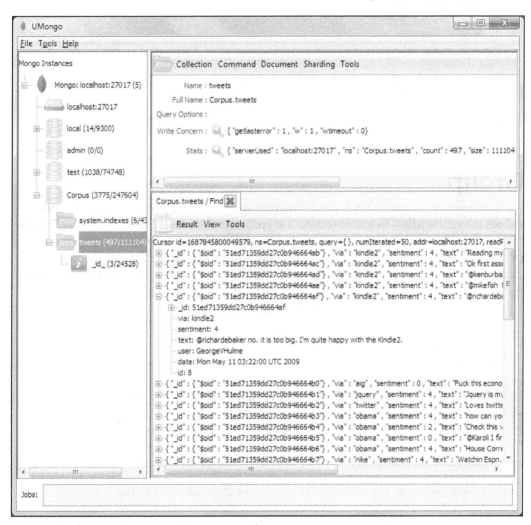

 All the codes and datasets of this chapter may be found in the author's GitHub repository at `https://github.com/hmcuesta/PDA_Book/tree/master/Chapter12`.

The complete code is as follows:

```
import json
from pymongo import MongoClient
con = MongoClient()
db = con.Corpus
tweets = db.tweets

with open("test.txt") as f:
  data = json.loads(f.read())
  for tweet in data["rows"]:
    tweets.insert(tweet)
```

Group

An aggregation function is a type of function used in data processing by grouping values into categories, in order to find a significant meaning. The common aggregate functions include **count**, **average**, **maximum**, **minimum**, and **sum**. However, we may perform more complicated statistical functions such as **mode** or **standard deviation**. Typically the grouping is performed with the SQL GROUP BY statement, as shown in the following code, additionally we may use aggregation functions such as COUNT, MAX, MIN, SUM in order to retrieve summarized information:

```
SELECT sentiment, COUNT(*)
FROM Tweets
GROUP BY sentiment
```

In MongoDB, we may use the group function, which is similar to SQL Group By statement. However, the group function doesn't work in shared systems and the result size is limited to 10,000 documents (20,000 in the version 2.2 or newer). Due to this, the group function is not highly used. Nevertheless, it is an easy way to find aggregate information when we have only one MongoDB instance.

In the following code, we may see the group function applied to the collection. The group command needs a key which is a field or fields to be grouped. Then, we will define a reduce function which will implement the aggregation function, in this case, the count of documents grouped by the sentiment field. Finally we define the initial value for the aggregation result document:

```
db.collection.group({
  key:{sentiment:true},
  reduce: function(obj,prev{prev. sentimentsum += obj.c}),
  initial: {sentimentsum: 0}
});
```

In the following screenshot, we can see the `group` function to find the number of tweets by `sentiment` (polarity) using UMongo. First we need to left-click on the `Tweets` collection and select **Group**.

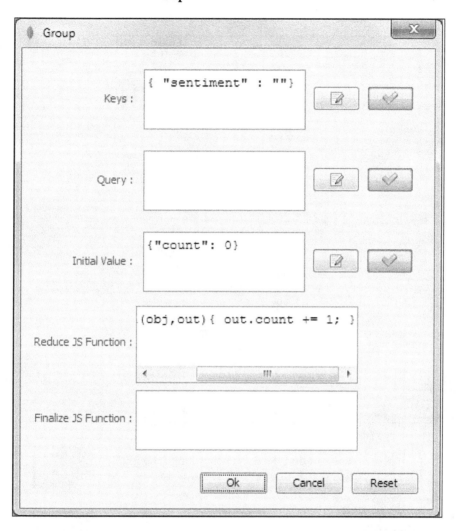

Then, we add {"sentiment" :""} in **Keys** and click on the **OK** button. The result can be seen in the following screenshot, bringing three categories with corresponding values for sentiments and count as 4.0 (or positive) and 181.0, 0.0 (or neutral) and 177.0, and 2.0 (or negative) and 139.0:

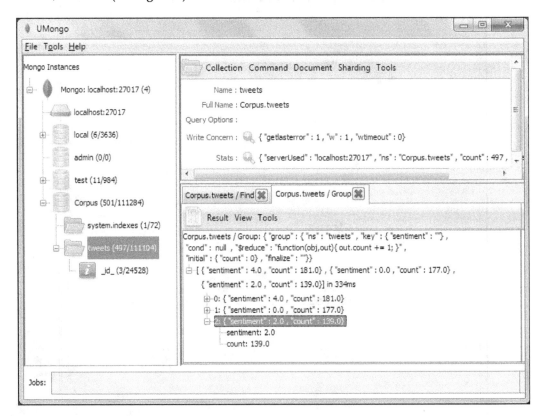

You can find the documentation of the group function at http://bit.ly/15iICc5.

In the following code, we can see how to perform a grouping in PyMongo using the group function of the tweets collection:

```
from pymongo import MongoClient
con = MongoClient()
db = con.Corpus
```

```
tweets = db.tweets

categories = tweets.group(key={"sentiment":1},
  condition={},
    initial={"count": 0},
      reduce="function(obj,   prev)
        {prev.count++;}")
for doc in categories:
  print(doc)
```

The result of the previous code will look as follows:

```
>>>
{'count': 181.0, 'sentiment': 4.0}
{'count': 177.0, 'sentiment': 0.0}
{'count': 139.0, 'sentiment': 2.0}
>>>
```

We may filter the result before the grouping with the cond attribute in mongo shell or conition in PyMongo. This is analogous to the WHERE statement in SQL:

```
cond: { via: "kindle2" },
```

The group function in PyMongo will look like as follows:

```
tweets.group(key={"sentiment":1},
  condition={"via": "kindle2" },
  initial={"count": 0},
  reduce="function(obj,   prev)
    {prev.count++;}")
```

The aggregation framework

The MongoDB aggregation framework is an easy way to get aggregated values and works well with **sharding** without having to use MapReduce (see *Chapter 13, Working with MapReduce*). The aggregation framework is flexible, functional, and simple to implement operational pipelines and computational expressions. The aggregation framework uses a declarative JSON format implemented in C++ instead of JavaScript, which improve the performance. The aggregate method prototype is shown as follows:

```
db.collection.aggregate( [<pipeline>] )
```

In the following code, we can see a simple counting by grouping the `sentiment` field with the `aggregate` method. In this case, the pipeline is only using the `$group` operator:

```
from pymongo import MongoClientcon = MongoClient()
db = con.Corpus
tweets = db.tweets

results = tweets.aggregate([
  {"$group": {"_id": "$sentiment", "count": {"$sum": 1}}}
  ])

for doc in results["result"]:
  print(doc)
```

In the following screenshot, we can see the result of the aggregation by grouping:

```
>>> ================================ RESTART ====================================
>>>
{'count': 139, '_id': 2}
{'count': 177, '_id': 0}
{'count': 181, '_id': 4}
>>>
```

You can find the documentation of the aggregation framework at
`http://docs.mongodb.org/manual/reference/aggregation/`.

Pipelines

In a pipeline we will process a stream of documents where the original input is a collection and the final output is a result document. The pipeline has a series of operators that filter or transform data, and generate a new document or filter out a document.

The following are the main pipeline operators:

- `$match`: It filters documents, uses existing query syntax and no geospatial operations or `$where`
- `$group`: It groups documents by an id and can use all the computational expressions such as `$max`, `$min`, and so on

- $unwind: It operates on an array field, yield documents for each array value, and also complements $match and $group

- $sort: It sorts documents by one or more fields

- $skip: It skips over documents in the pipeline

- $limit: It restricts the number of documents in an aggregation pipeline

In the following code we can see the aggregation with a pipeline using the $group, $sort, and $limit operators:

```
from pymongo import MongoClientcon = MongoClient()
db = con.Corpus
tweets = db.tweets

results = tweets.aggregate([
  {"$group": {"_id": "$via",
                      "count": {"$sum": 1}}},
  {"$sort": {"via":1}},
  {"$limit":10},
  ])

for doc in results["result"]:
  print(doc)
```

In the following screenshot we can see the result of the aggregation using a pipeline with multiple operators:

```
>>> ================================ RESTART ================================
>>>
{'count': 1, '_id': 'fred wilson'}
{'count': 8, '_id': 'warren buffet'}
{'count': 1, '_id': 'aapl'}
{'count': 2, '_id': 'mashable'}
{'count': 1, '_id': 'hitler'}
{'count': 1, '_id': 'yankees'}
{'count': 1, '_id': 'republican'}
{'count': 7, '_id': 'exam'}
{'count': 1, '_id': 'world cup'}
{'count': 5, '_id': 'viral marketing'}
>>>
```

Expressions

The expressions produce output documents based on calculations performed on input documents. The expressions are stateless and are only used in the aggregation process.

The $group aggregation operations are:

- $max: It return the highest value in the group
- $min: It return the lowest value in the group
- $avg: It return the average of all the group values
- $sum: It return the sum of all values in the group
- $addToSet: It returns an array of all the distinct values for a certain field in each document in that group

We can also find other kinds of operators depending on its data type as follows:

- **Boolean**: $and, $or, and $not
- **Arithmetic**: $add, $divide, $mod, $multiply, and $substract
- **String**: $concat, $substr, $toUpper, $toLower, and $strcasecmp
- **Conditional**: $cond and $ifNull

In the following code, we will use the aggregate method with the $group operator and in this case we will use multiple operations such as $avg, $max, and $min:

```
from pymongo import MongoClient
con = MongoClient()
db = con.Corpus
tweets = db.tweets

results = tweets.aggregate([
  {"$group": {"_id": "$via",
    "avgId": {"$avg": "$id"} ,
    "maxId": {"$max": "$id"} ,
    "minId": {"$min": "$id"} ,
    "count": {"$sum": 1}}}
  ])
for doc in results["result"]:
  print(doc)
```

In the following screenshot, we can see the result of $group using multiple operators:

```
>>> ================================ RESTART ================================
>>>
{'count': 7, 'avgId': 1065.857142857143, '_id': 'exam', 'maxId': 2195, 'minId': 218}
{'count': 1, 'avgId': 226.0, '_id': 'republican', 'maxId': 226, 'minId': 226}
{'count': 1, 'avgId': 1025.0, '_id': 'world cup', 'maxId': 1025, 'minId': 1025}
{'count': 1, 'avgId': 2398.0, '_id': 'yankees', 'maxId': 2398, 'minId': 2398}
{'count': 1, 'avgId': 14045.0, '_id': 'aapl', 'maxId': 14045, 'minId': 14045}
{'count': 1, 'avgId': 2296.0, '_id': 'hitler', 'maxId': 2296, 'minId': 2296}
...
```

The aggregation framework has some limitations such as the document size limit is 16MB and there are some field types unsupported (Binary, Code, MinKey and MaxKey).

In terms of sharding support sharding support, MongoDB analyses pipeline and forwards operations up to $group or $sort to shards, then combines shard server result and returns them. Due to this, it is recommended to use $match and $sort as early as possible into the pipeline.

Summary

In this chapter, we explored the basic operations and functions of MongoDB. We also performed a data preparation of a CSV dataset with OpenRefine and turned it into a well-formatted JSON dataset. Finally, we present a data processing introduction with the aggregation framework, which is a faster alternative to MapReduce for common aggregations. We introduced the basic operators used in the pipelines and the expressions supported by the aggregation framework.

In the next chapter, we will explore the MapReduce functionality of MongoDB and we will create a word-cloud in D3 with the most frequent words in positive tweets.

13
Working with MapReduce

MongoDB is a document-based database used to tackle large amounts of data and is used by companies such as Forbes, Bitly, Foursquare, Craigslist, and so on. In *Chapter 12, Data Processing and Aggregation with MongoDB*, we learned how to perform the basic operations and aggregations with MongoDB. In this chapter, we will learn how MongoDB implements a MapReduce programming model.

In this chapter we will cover:

- MapReduce overview
- Programming model
- Using MapReduce with MongoDB
  - ° The `map` function
  - ° The `reduce` function
  - ° Using mongo shell
  - ° Using UMongo
  - ° Using PyMongo
- Filtering the input collection
- Grouping and aggregation
- The most common words in tweets in a word-cloud visualization

You can find a list of production deployments of MongoDB at
`http://www.mongodb.org/about/production-deployments/`.

MapReduce overview

MapReduce is a programming model for large-scale distributed data processing. It is inspired by the `map` function and the `reduce` function of the functional programming languages such as Lisp, Haskell, or Python. One of the most important features of MapReduce is that it allows us to hide the low-level implementation such as message passing or synchronization from users and allows to split a problem into many partitions. This is a great way to make trivial parallelization of data processing without any need for communication between the partitions.

> Google's original paper: *MapReduce: Simplified Data Processing on Large Clusters*, can be found at `http://research.google.com/archive/mapreduce.html`.

MapReduce became main stream because of Apache Hadoop, which is an open source framework that was derived from Google's MapReduce paper. MapReduce allows us to process massive amounts of data in a distributed cluster. In fact, there are many implementations of the MapReduce programming model. Some of them are shown in the following list. It is important to say that MapReduce is not an algorithm; it is just a part of a high-performance infrastructure that provides a lightweight way to run a program in a lot of parallel machines.

Some of the most popular implementations of MapReduce are listed as follows:

- **Apache Hadoop**: It is probably the most famous implementation of Google's MapReduce model, based on Java with an excellent community and vast ecosystem. We can find more information about it at `http://hadoop.apache.org/`.

- **MongoDB**: It is a document-oriented database which provides MapReduce operations. We can find more information about it at `http://docs.mongodb.org/manual/core/map-reduce/`.

- **Phoenix system**: It is a Google's MapReduce implementation that can be used in multi-core and shared-memory multiprocessors, originally created as a class project at Stanford. We can find more information about it at `http://mapreduce.stanford.edu/`.

- **MapReduce-MPI library**: It is a MapReduce implementation which runs on top of **MPI (Message Passing Interface)** standard. You can find more information about it at `http://mapreduce.sandia.gov/`.

> Message passing is a technique used in concurrent programming to provide synchronization among processes, similar to a traffic light control system. MPI is a standard for message passing implementation. We can find more information about MPI at http://en.wikipedia.org/wiki/Message_Passing_Interface.

Programming model

MapReduce provides an easy way to create parallel programs without the concern for message passing or synchronization. This can help us to perform complex aggregation tasks or searches. As we can observe in the following figure, MapReduce can work with less organized data (such as noise, text, or schemaless documents) than the traditional relational databases. However, the programming model is more procedural which means that the user must have some programming skills such as Java, Python, JavaScript, or C. MapReduce requires two functions, the map function which is going to create a list of key-value pairs and the reduce function, which will iterate over each value and then apply a process (merge or summarization) to get an output.

In MapReduce, the data could be split into several nodes (sharding) in that case we will need a partition function. The partition function will be in charge of sort and load balancing. In MongoDB we can work over sharded collections automatically without any configuration.

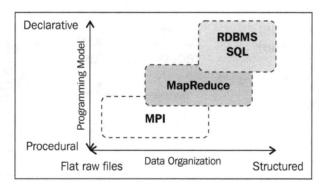

Using MapReduce with MongoDB

MongoDB provides us with a `mapReduce` command and in the following figure, we can observe the life circle of the MapReduce process in MongoDB. We start with a **Collection** or a query and each document in the collection will call the `map` function. Then, using the `emit` function we will create an intermediate hash map (See the following figure) with a list of pairs (key-value). Next, the `reduce` function will iterate the intermediate hash map and it will apply some operation to all values of each key. Finally, the process will create a brand new collection with the output. The map/reduce functions in MongoDB will be programmed with JavaScript.

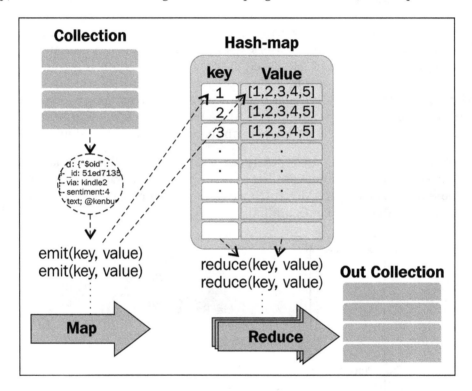

 You can find the reference documentation of MapReduce with MongoDB at http://docs.mongodb.org/manual/core/map-reduce/.

The map function

The map function will call the emit function one or more times (See the next figure). We can access all the attributes of each document in the collection with the this keyword. The intermediate hash map contains only unique keys, so if the emit function sends a key that is already in the hash map, that value is going to be inserted in a list of values. Each record in the hash map will look similar to: key:One, value:[1,2,3,…].

The following code is a sample code of the map function:

```
function(){
    emit(this._id, {count: 1});
}
```

The reduce function

The reduce function will receive two arguments, key and values (one value or a list of values). This function is going to be called for each record in the hash map.

In the following code we can see a sample reduce function. In this case the function is going to return the total count for each key.

```
function(key, values) {
    total = 0;
    for (var i = 0; i < values.length; ++i) {
        total += values[i].count;
    };
    return {count: total};
}
```

Refer the next section to see the reduce function's working.

Using mongo shell

Mongo shell provides a wrapper method for the mapReduce command. The db.collection.mapReduce() method must receive three parameters, the map function, the reduce function, and the name of the collection where the output is going to be stored, as is shown in the following command. Refer to the *Installing and running MongoDB* section in *Appendix, Setting Up the Infrastructure*, to find complete instructions on how to install and run MongoDB and mongo shell.

```
db.collection.mapReduce(map,reduce,{out:"OutCollection"})
```

In this example we will use the `tweets` collection that we already created in the *Inserting Documents with PyMongo* section of *Chapter 12, Data Processing and Aggregation with MongoDB*, with the attributes id, via, sentiment, text, user, and date. This example will count how many times each unique element of the via attribute appears in the collection.

First, we need to define the map function in the `mapTest` variable:

```
mapTest = function(){
  emit(this.via, 1);
  }
```

Then, we need to define the reduce function in the `reduceTest` variable:

```
reduceTest = function(key, values) {
  var res = 0;
    values.forEach(function(v){ res += 1})
  return {count: res};
  }
```

The mongo shell will look similar to the following screenshot:

```
> mapTest = function(){ emit(this.via, 1); }
function (){ emit(this.via, 1); }
> reduceTest = function(key, values) {
...
... var res = 0;
... values.forEach(function(v){ res += 1})
...
... return {count: res};
... }
function (key, values) {

var res = 0;
values.forEach(function(v){ res += 1})

return {count: res};
}
>
```

Now, we need to define Corpus as the default database:

```
use Corpus
```

Next, we will use the `mapReduce` method to send the `mapTest` function and the `reduceTest` function, and for defining a new collection results to store the output:

```
db.tweets.mapReduce(mapTest,reduceTest,{out:"results"})
```

Finally, we will retrieve all the documents of the `results` collection with the `find` method:

```
db.results.find()
```

In the following screenshot we can see the result of the `mapReduce` command in the mongo shell and the retrieved collection (`results`) with the aggregated data (`count`) of the `via` attribute.

```
> use Corpus
switched to db Corpus
> db.tweets.mapReduce(mapTest,reduceTest, {out:"results"})
{
        "result" : "results",
        "timeMillis" : 135,
        "counts" : {
                "input" : 497,
                "emit" : 497,
                "reduce" : 59,
                "output" : 80
        },
        "ok" : 1,
}
> db.results.find()
{ "_id" : 40, "value" : { "count" : 4 } }
{ "_id" : 50, "value" : { "count" : 6 } }
{ "_id" : "Bobby Flay", "value" : { "count" : 8 } }
{ "_id" : "Danny Gokey", "value" : { "count" : 4 } }
{ "_id" : "Malcolm Gladwell", "value" : { "count" : 11 } }
{ "_id" : "aapl", "value" : 1 }
{ "_id" : "aig", "value" : { "count" : 7 } }
{ "_id" : "at&t", "value" : { "count" : 15 } }
{ "_id" : "bailout", "value" : 1 }
{ "_id" : "baseball", "value" : { "count" : 6 } }
{ "_id" : "bing", "value" : 1 }
{ "_id" : "booz allen", "value" : { "count" : 3 } }
{ "_id" : "car warranty call", "value" : { "count" : 2 } }
{ "_id" : "cheney", "value" : { "count" : 5 } }
{ "_id" : "china", "value" : { "count" : 6 } }
{ "_id" : "comcast", "value" : { "count" : 4 } }
{ "_id" : "dentist", "value" : { "count" : 17 } }
{ "_id" : "driving", "value" : 1 }
{ "_id" : "east palo alto", "value" : { "count" : 4 } }
{ "_id" : "eating", "value" : { "count" : 12 } }
Type "it" for more
>
```

 For complete reference of the `mapReduce` command we can follow the link http://bit.ly/13Yh5Kg.

Using UMongo

In *Chapter 12, Data Processing and Aggregation with MongoDB*, we have learned how to use UMongo to perform queries and grouping. In this section, we will use UMongo to execute a `mapReduce` command from a user interface. First, we will open and connect UMongo with the local MongoDB. Refer to the *Installing and running UMongo* section in *Appendix, SettingUp the Infrastructure*, for complete instructions on how to install and run UMongo.

Now, we will select the **Corpus** database and the **tweets** collection, then, right-click on **tweets** to display the options.

Next, as we can see in the following screenshot, we will click on the **Map Reduce** option:

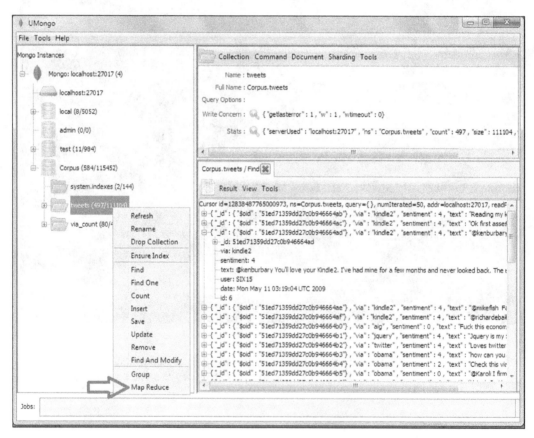

Then, we will see the **Map Reduce** window, where we can insert the **Map JS Function** and the **Reduce JS Function** text areas and also define the **Input** filtering and the **Output** collection. In the following screenshot, we can see the functions to count the number of occurrences of each unique element of the via attribute in the collection. However, this time we are filtering the input and only the positive tweets are considered. While inserting {sentiment: 4} in the query text area, we will consider only the documents where the attribute sentiment is equal to 4 (2 = negatives and 4 = positives). Refer to the *Filtering the input collection* section, to find about the details of the query and its operators. Finally, in the **Output Collection** field, we write via_count, which is the collection where the output is going be stored.

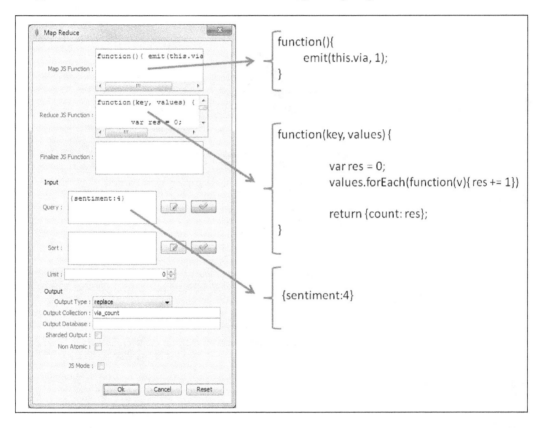

Finally, to check the result, we will select the collection **via_count** and left-click on the **Find** method to see all the documents created by the `mapReduce` command. In the following screenshot we can see the result. Each document in the **via_count** collection will look similar to the following code snippet:

```
{"id":"google", "values":{"count":3}}
```

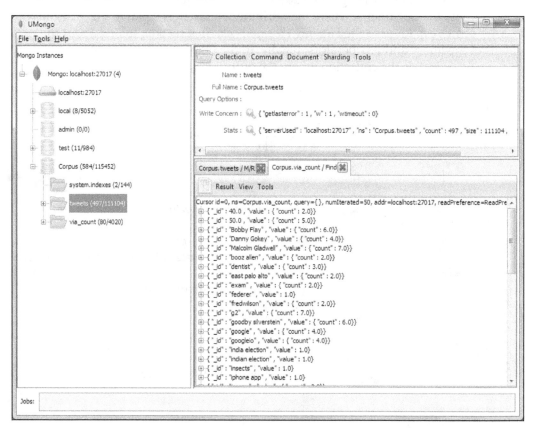

Using PyMongo

With mongo shell or UMongo we can run MapReduce process in an easy way. However, we will normally need to use the MapReduce process as a part of a bigger transaction. Then, we need to implement a MapReduce wrapper in an external programming language. In this case, we will use PyMongo to call the `mapReduce` command from Python.

In this example, we will use the `tweets` collection and we will count how many times the `via` attribute appears. Refer to the *Inserting Documents with PyMongo* section in *Chapter 12, Data Processing and Aggregation with MongoDB*, for details about the creation of the `tweets` collection.

First, we will import the `pymongo` and `bson.code` modules. Refer to the *Installing and running MongoDB* section in *Appendix, Setting Up the Infrastructure*, to find instructions on how to install PyMongo:

```
from pymongo import MongoClientfrom bson.code import Code
```

Then, we will establish connection with the MongoDB service to the default localhost and port 27017.

```
con = MongoClient()
```

Next, we will define `Corpus` as the default database and `tweets` as a shortcut object of `db.tweets`:

```
db = con.Corpus
tweets = db.tweets
```

Now, we will use the object constructor `Code` for representing JavaScript functions, `map` and `reduce` in BSON, since the MongoDB API methods use JavaScript.

```
map = Code("function(){ emit(this.via, 1); }")

reduce = Code("""function(key, values) {
  var res = 0;
  values.forEach(function(v){ res += 1})
  return {count: res};
}""")
```

Then, we will use the `map_reduce` function providing three parameters, the `map` function, the `reduce` function, and define `via_count` as the output collection.

```
result = tweets.map_reduce(map,reduce,"via_count")
print(result)
```

Finally, we retrieve all the documents in the `via_count` collection with the `find` function.

```
for doc in db.via_count.find():
  print(doc)
```

In the following screenshot, we can observe the result of this code in the IDLE:

```
7% Python Shell                                                    □ □ X
File  Edit  Shell  Debug  Options  Windows  Help
>>> ================================ RESTART ========================
>>>
{'counts': {'input': 497, 'reduce': 59, 'emit': 497,
 'output': 80}, 'timeMillis': 55, 'ok': 1.0, 'result': 'via count'}

{'_id': 40.0, 'value': {'count': 4.0}}
{'_id': 50.0, 'value': {'count': 6.0}}
{'_id': 'Bobby Flay', 'value': {'count': 8.0}}
{'_id': 'Danny Gokey', 'value': {'count': 4.0}}
{'_id': 'Malcolm Gladwell', 'value': {'count': 11.0}}
{'_id': 'aapl', 'value': 1.0}
{'_id': 'aig', 'value': {'count': 7.0}}
{'_id': 'at&t', 'value': {'count': 15.0}}
{'_id': 'bailout', 'value': 1.0}
{'_id': 'baseball', 'value': {'count': 6.0}}
{'_id': 'bing', 'value': 1.0}
{'_id': 'booz allen', 'value': {'count': 3.0}}
{'_id': 'car warranty call', 'value': {'count': 2.0}}
                                                    Ln: 3393 Col: 4
```

Filtering the input collection

Sometimes, we don't need the entire collection for our MapReduce process. Hence, the mapReduce command provides us with optional parameters to filter the input collection.

The parameter query allows us to apply criteria using the query operators to filter the document's input to the map function. In the following code, we will filter the documents in the collection, and only include the documents where the attribute number is greater than 10 ("$gt":10):

```
collection.map_reduce(map_function,
    reduce_function,
    "output_collection",
    query={"number":{"$gt":10}})
```

The query operators used in the MapReduce `query` parameter are the same query selectors seen in the *Getting started with MongoDB* section in *Chapter 12, Data Processing and Aggregation with MongoDB*, used to perform simple queries. In the following table we present the most common operators and their equivalent in SQL language:

Mongo operators	SQL operators
$gt	>
$gte	>=
$in	IN
$lt	<
$lte	<=
$and	AND
$or	OR

The `limit` parameter is an optional parameter of the `mapReduce` command, which helps us to define the maximum number of documents retrieved by the query. In the following code we define the limit of documents retrieved to a maximum of `10`:

```
collection.map_reduce(map_function,
   reduce_function,
   "output_collection",
   limit = 10)
```

 You can find a complete list of MongoDB operators at
`http://docs.mongodb.org/manual/reference/operator/`.

Grouping and aggregation

In the following example we will perform grouping and aggregation in order to get statistics (`sum`, `max`, `min`, and `avg`) about NBA players and their number of points scored. First, the `map` function will send the name of the `player` and the number of `points` scored for each game. The `map` function will look similar to the following code:

```
function(){emit(this.player, this.points); }
```

Then, we can perform all the aggregation functions simultaneously using the method `sum` from the JavaScript `Array` object and the `max`/`min` functions of the JavaScript `Math` object. The `reduce` function will look similar to the following code:

```
function(key, values) {
  var explain = {total:Array.sum(values),
    max:Math.max.apply(Math, values),
```

```
      min:Math.min.apply(Math, values),
      avg:Array.sum(values)/values.length}
    return explain
}
```

For this example we will create synthetic data, randomly mixing the name of the 10 players and we will assign a random score between 0 and 100. Then, we will insert the data into a MongoDB collection called Games. The complete code is listed as follows:

```
import random as ran
import pymongo
con = pymongo.Connection()
db = con.basketball
games = db.games

players = ["LeBron James",
    "Allen Iverson",
    "Kobe Bryant",
    "Rick Barry",
    "Dominique Wilkins",
    "George Gervin",
    "Dwyane Wade",
    "Jerry West",
    "Pete Maravich",
    "Carmelo Anthony"]
for x in range(100):
    games.insert({ "player" : players[ran.randint(0,9)],
    "points" : ran.randint(0,100)})
```

The collection Games will look similar to the following screenshot. We can observe that a player can appear several times with different point scores:

```
{ "_id" : { "$oid" : "5206caef9dd27c1964b1d648"} , "player" : "LeBron James" , "points" : 40}
{ "_id" : { "$oid" : "5206caef9dd27c1964b1d649"} , "player" : "Rick Barry" , "points" : 6}
{ "_id" : { "$oid" : "5206caef9dd27c1964b1d64a"} , "player" : "George Gervin" , "points" : 0}
{ "_id" : { "$oid" : "5206caef9dd27c1964b1d64b"} , "player" : "Kobe Bryant" , "points" : 56}
{ "_id" : { "$oid" : "5206caef9dd27c1964b1d64c"} , "player" : "Pete Maravich" , "points" : 4}
{ "_id" : { "$oid" : "5206caef9dd27c1964b1d64d"} , "player" : "Dwyane Wade" , "points" : 65}
{ "_id" : { "$oid" : "5206caef9dd27c1964b1d64e"} , "player" : "Pete Maravich" , "points" : 55}
{ "_id" : { "$oid" : "5206caef9dd27c1964b1d64f"} , "player" : "Dwyane Wade" , "points" : 45}
{ "_id" : { "$oid" : "5206caef9dd27c1964b1d650"} , "player" : "Allen Iverson" , "points" : 66}
{ "_id" : { "$oid" : "5206caef9dd27c1964b1d651"} , "player" : "Rick Barry" , "points" : 18}

. . .
```

Finally, we will perform the MapReduce process using the `map`/`reduce` functions seen at the beginning of this section and implement it in `pymongo`. We will store the output in the `_result` collection. We can see the complete code as follows:

```
from pymongo import MongoClient from bson.code import Code
con = MongoClient()db = con.basketball
games = db.games

map = Code("""function(){
    emit(this.player, this.points);
  }""")

reduce = Code("""function(key, values) {
   var explain = {total:Array.sum(values),
     max:Math.max.apply(Math, values),
     min:Math.min.apply(Math, values),
     avg:Array.sum(values)/values.length}
     return explain;
   }""")

result = games.map_reduce(map,reduce,"_result")
print(result)
```

The result of the grouping and aggregation will look similar to the following screenshot:

```
>>> ============================= RESTART =============================
>>>
Collection(Database(Connection('localhost', 27017), 'baseball'), '_result')
{'_id': 'Allen Iverson', 'value': {'max': 66.0, 'total': 310.0, 'avg': 34.44444444444444, 'min': 9.0}}
{'_id': 'Carmelo Anthony', 'value': {'max': 91.0, 'total': 473.0, 'avg': 47.3, 'min': 1.0}}
{'_id': 'Dominique Wilkins', 'value': {'max': 98.0, 'total': 545.0, 'avg': 60.55555555555556, 'min': 20.0}}
{'_id': 'Dwyane Wade', 'value': {'max': 95.0, 'total': 834.0, 'avg': 55.6, 'min': 15.0}}
{'_id': 'George Gervin', 'value': {'max': 81.0, 'total': 235.0, 'avg': 47.0, 'min': 0.0}}
{'_id': 'Jerry West', 'value': {'max': 98.0, 'total': 645.0, 'avg': 58.63636363636363, 'min': 9.0}}
{'_id': 'Kobe Bryant', 'value': {'max': 95.0, 'total': 497.0, 'avg': 45.18181818181818, 'min': 0.0}}
{'_id': 'LeBron James', 'value': {'max': 100.0, 'total': 546.0, 'avg': 49.63636363636363, 'min': 3.0}}
{'_id': 'Pete Maravich', 'value': {'max': 97.0, 'total': 562.0, 'avg': 43.23076923076923, 'min': 4.0}}
{'_id': 'Rick Barry', 'value': {'max': 98.0, 'total': 781.0, 'avg': 48.8125, 'min': 6.0}}
>>> |
```

 All the codes and datasets of this chapter can be found in the author's GitHub repository at `https://github.com/hmcuesta/PDA_Book/tree/master/Chapter13`.

Word cloud visualization of the most common positive words in tweets

In this example, we will develop a simple application that counts the number of occurrences of each word in the positive tweets. First, we will split each tweet into words. Then, we remove all the URLs (http://...) and twitter users (@...). Next, we will remove all the words with three or less characters (such as the, why, she, him, and so on). Finally, the counted word frequencies will be visualized into a word cloud. In the code listed as follows, we implement the JavaScript map function to split words from tweets:

```
function(){
  this.text.split(' ').forEach(
     function(word){
       var txt = word.toLowerCase();
         if(!(/^@/).test(txt) &&
            txt.length >= 3 &&
          !(/^http/).test(txt)){
       emit(txt,1)
       }
     }
  }
}
```

The input will look similar to the following code snippet:

```
'text': '@SomeUsr After using LaTeX a lot any other typeset
mathematics just looks greate. http://www.latex.org',
```

The output will look similar to the following code snippet. For each word, the emit function will be called:

```
["after", "using", "latex",  "other", "typeset", "mathematics", "
just", "looks", "great"]
```

In the code listed as follows, we implement the JavaScript reduce function to get the frequency of occurrence of each word:

```
function(key, values) {
  var res = 0;
  values.forEach(function(v){ res += 1})
  return {count: res};
}
```

> In *Chapter 11, Sentiment Analysis of Twitter Data*, we already discussed how a bag-of-words model is a common method for document classification by using the frequency of occurrences of each word as a feature for the classifier.

For this example we will use the database `Corpus` and the `tweets` collection created in the *Inserting Documents with PyMongo* section in *Chapter 12, Data Processing and Aggregation with MongoDB*. Each document in the `tweets` collection will look similar to the following format:

```
{'via': 'latex',
 'sentiment': 4,
 'text': '@SomeUsr After using LaTeX a lot any other typeset
mathematics just looks greate. http://www.latex.org',
 'user': 'yomcat',
 'date': 'Sun Jun 14 04:31:28 UTC 2009',
 '_id': ObjectId('51ed71359dd27c0b94666696'),
 'id': 14071}
```

In the following code, we will implement a `map_reduce` method for querying only the positive tweets (`sentiment` = 4) as an input collection:

```
from pymongo import MongoClientfrom bson.code import Code
import csv

con = MongoClient()db = con.Corpus
tweets = db.tweets
map = Code("""function(){
  this.text.split(' ').forEach(
    function(word){
    var txt = word.toLowerCase();
      if(!(/^@/).test(txt) &&
        txt.length > 3 &&
      !(/^http/).test(txt)){
    emit(txt,1)
    }
  }
  )
}""")

reduce = Code("""function(key, values) {
    var res = 0;
    values.forEach(function(v){ res += 1})
    return {count: res};
    }""")

result = tweets.map_reduce(map,reduce,"TweetWords",
query={"sentiment":4})
```

The output collection will be stored in `TweetWords`. We can check the number of resulted words (2173) with the following command:

```
db.runCommand( { count: TweetWords })
```

In the following screenshot, we can see the count and content of the collection `TweetWords`:

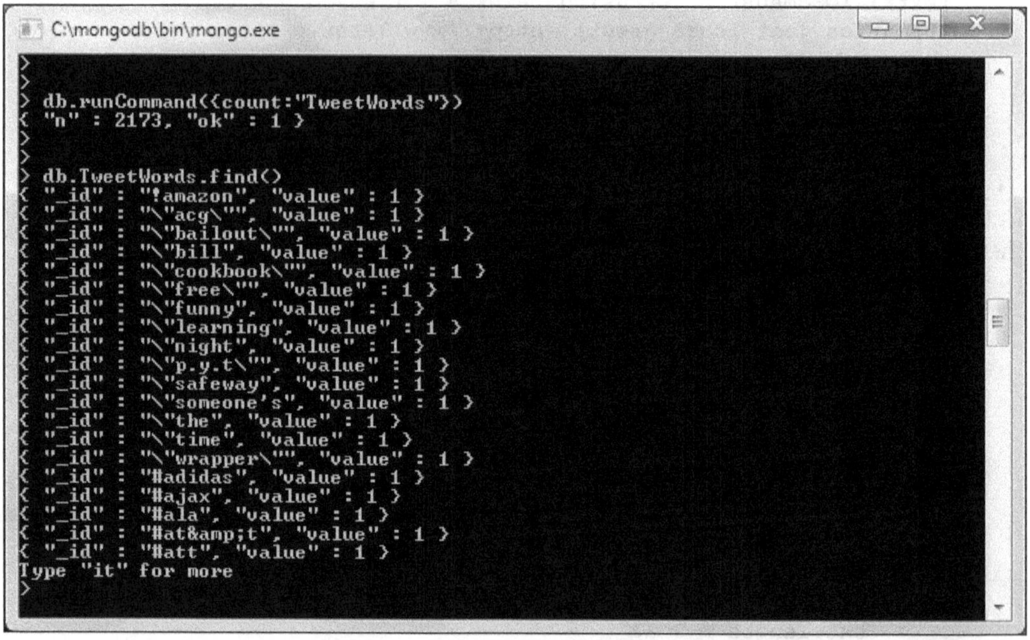

Now, for our visualization we need a `csv` file with the 50 most frequent words. In the following code we will perform a query on the `TweetWords` collection by sorting the result in descending order and limiting the output to only the first 50 documents. Finally, we will store the output in the file `data.csv`:

```
with open("data.csv", "w") as f:
    f_csv = csv.writer(f, delimiter=',')
    f_csv.writerow(["text","size"])

    for doc in db.TweetWords.find()
      .sort("value", direction = -1)
      .limit(50):
        f_csv.writerow([doc["_id"],doc["value"]["count"]+30])
        print(doc)
```

We can see the output of the query in the following screenshot:

```
76 Python Shell                                                    [ - ][ □ ][ X ]

File  Edit  Shell  Debug  Options  Windows  Help
>>>
Collection(Database(Connection('localhost', 27017), 'Corpus'), 'TweetWords')
{'_id': 'love', 'value': {'count': 28.0}}
{'_id': 'good', 'value': {'count': 18.0}}
{'_id': 'just', 'value': {'count': 18.0}}
{'_id': 'with', 'value': {'count': 18.0}}
{'_id': 'have', 'value': {'count': 17.0}}
{'_id': 'night', 'value': {'count': 15.0}}
{'_id': 'from', 'value': {'count': 13.0}}
{'_id': 'nike', 'value': {'count': 13.0}}
                                                            Ln: 2882 Col: 4
```

In this example we will use the word-cloud layout written in D3.js by Jason Davies.
We will obtain data.csv created in the earlier python code using the d3.csv function.

> You can download the d3-cloud layout from the Jason Davies GitHub
> repository at https://github.com/jasondavies/d3-cloud/.

In the following code we can see the implementation of the d3.layout.cloud.js
file using the data obtained from MongoDB:

```html
<!DOCTYPE html>
<meta charset="utf-8">
<body>
<script src="http://d3js.org/d3.v3.min.js"></script>
<script src="d3.layout.cloud.js"></script>
<script>
  var fill = d3.scale.category20();
  var data = [];

  d3.csv("data.csv", function(w) {
    w.forEach( function (d) {
        data.push({text: d.text, size: d.size});
    });
```

The data.csv file will look similar to the following file:

```
text,size
love,58.0
good,48.0
just,48.0
```

```
with,48.0
have,47.0
night,45.0
from,43.0
nike,43.0
.  .  .


  d3.layout.cloud().size([800, 400])
    .words(data)
    .padding(5)
    .rotate(function() { return ~~(Math.random() * 2) * 90; })
    .font("Impact")
    .fontSize(function(d) { return d.size; })
    .on("end", draw)
    .start();
});

  function draw(words) {
    d3.select("body").append("svg")
        .attr("width", 800)
        .attr("height", 400)
      .append("g")
        .attr("transform", "translate(350,250)")
      .selectAll("text")
        .data(words)
      .enter().append("text")
        .style("font-size", function(d) { return d.size + "px"; })
        .style("font-family", "Impact")
        .style("fill", function(d, i) { return fill(i); })
        .attr("text-anchor", "middle")
        .attr("transform", function(d) {
          return "translate(" + [d.x, d.y] + ")rotate(" + d.rotate +
")";
        })
        .text(function(d) { return d.text; });
  }
</script>
```

In the following screenshot we can see the result in the word cloud of the most frequent words in the positive tweets:

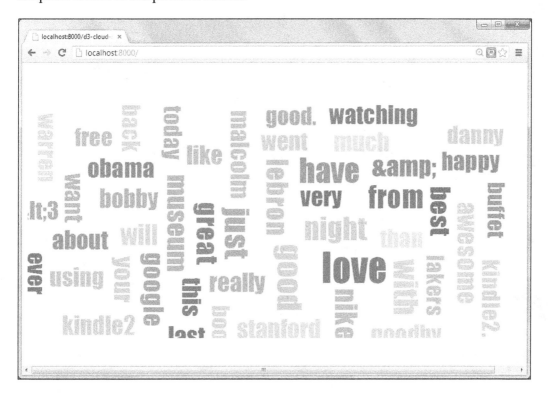

Summary

In this chapter, we explored the basic concepts of the MapReduce programming model and how to implement common activities such as grouping, aggregation, counting, and summing in MongoDB.

MapReduce is a powerful tool for log analysis and data processing. In this chapter, we learned how to get easy but powerful aggregation capabilities implemented in Python using PyMongo.

In the next chapter, we will explore an online Python tool for data analysis and development called **Wakari**.

14
Online Data Analysis with IPython and Wakari

In this chapter, we will introduce an online tool for data analysis called **Wakari**, in which we will set up a complete Python environment within a few seconds. Then, we will present some of the capabilities of Wakari through IPython Notebook, by using the PIL and Pandas libraries.

In this chapter, we will cover:

- Getting started with Wakari
- Getting started with IPython Notebook:
    - Data Visualization

- Introduction to image processing with PIL:
    - Image object
    - Image histogram
    - Image filtering, operations, and transformations

- Introduction to data analysis with Pandas:
    - Working with time series
    - Working with multivariate dataset with the `DataFrame` object
    - Grouping, aggregation, and correlation

- Multiprocessing with IPython
- Sharing your notebook

Getting started with Wakari

Wakari is a cloud service for collaborative Python data-analysis environments, created by **Continuum Analytics**. Wakari provides a powerful set of preconfigured Python environments built over **Anaconda**, which is a free Python distribution for large-scale data processing and scientific computing. Wakari uses an IPython GUI, which is a Python shell improved for writing, debugging, and testing Python code for scientific computing. IPython provides a terminal-based interface and an HTML notebook similar to **Wolfram-Mathematica**. In Wakari, we can either use the terminal console or the IPython Notebook.

Wakari helps us to set up a complete scientific Python environment without any local installation. This can be very convenient for learning purposes, because we may start coding right away and the Anaconda distribution includes several of the most used libraries such, as NumPy, SciPy, Matplotlib, PIL, Pandas, Numba, and so on.

In Wakari, we may use different kinds of terminals such as Python, Shell, IPython, or SSH. However, in this chapter we will focus on the use of IPython Notebook.

IPython Notebook is a rich web interface for coding. The notebook is a great tool for teaching and presenting Python code in an interactive interface. In this chapter, we will use the IPython Notebook included in Wakari, and we will test some of their capabilities by implementing examples in **PIL (Python Image Library)** and Pandas.

> For more information about IPython, visit `http://ipython.org/`.

Creating an account in Wakari

To start working with Wakari, we need to create an account, or log in if we already have an account. We can create a new account using the following link:

`https://www.wakari.io/`

In the following screenshot, we can see the web form to register a new free account. In this chapter, we will work with the free account, which has some restrictions. However, we may find plans from 10 dollars, which can give us access via SSH and the capability to execute long-run jobs.

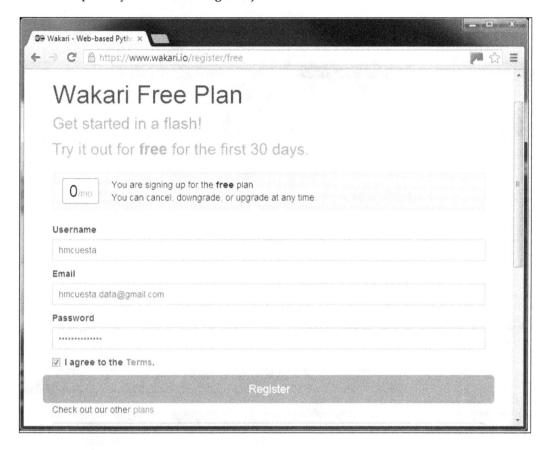

Once we log in to Wakari, the interface will look similar to the following screenshot, with tabs on the right side of the window for the terminals, IPython Notebooks and a **New Notebook** button. On the left side of the window, we may see the account path with the resources (files and folders uploaded by the user):

If we click on the **Terminals** tab, we may add a new Python shell, Linux shell, or IPython shell. In the following screenshot, we observe a new Python shell:

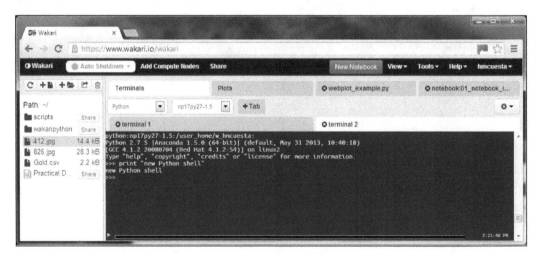

If we click on **Tools** and then select **Anaconda Environments**, we can see a complete list of the installed packages and modules as shown in the following screenshot:

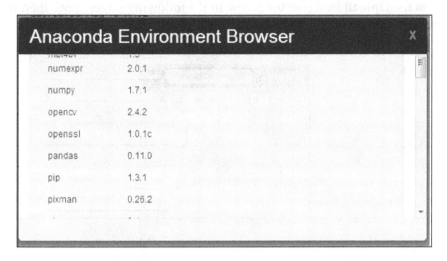

Getting started with IPython Notebook

The **IPython Notebook (NB)** is a web interface for our python code. NB is based in a JSON format, sharable and portable in `.pynb` file format.

To start with a blank notebook, we will click on the **New Notebook** button. In the following screenshot, we can see how to change the name by clicking on the **Untitled0** label, and then we will rename the notebook:

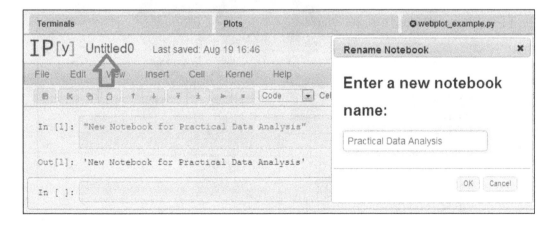

The NB will have access to all resources (text files, images, and so on) in the path. We can upload the text files, images, and other content to the Wakari platform by clicking on the **Upload** icon (see the arrow in the following screenshot), then we will select the files, and finally we will click on the **Upload Files** button as shown in the following screenshot:

Finally, we will click on the play icon (see the arrow in the following screenshot) to run the code of our NB. We will get a numbered output for each of our input codes as visible in the following screenshot. We may code several lines in the same input (In [1]) which we call cells, and as a result we can see the plot in the output (Out [1]). We also have access to all the modules included in the Anaconda distribution and all the resources in the path:

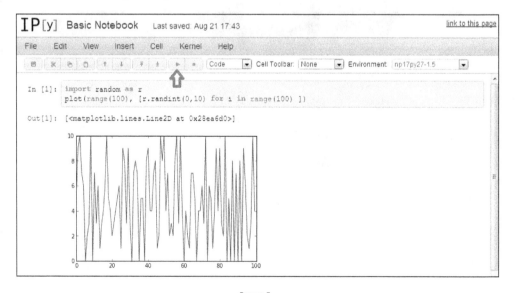

When we need to save the progress of our NB, we will click on the **File** menu and then select **Save**. If we need to create a local copy of our NB, we can click on the **File** menu and then click on **Download as**. Next, we can choose either the NB file (**.ipynb**) or the raw Python code (**.py**).

 You can find more information about IPython Notebook from http://ipython.org/notebook.html.

Data visualization

Wakari supports two methods of plotting. The first method is by using `matplotlib` and all its capabilities. PyLab is just a wrapper for modules such as `matplotlib`, `numpy`, and `scipy`, for numerical analysis and computation. In the following screenshot, we can see `plot_surface` implementing an `Axes3D` object:

```
In [2]:  from pylab import *
         from mpl_toolkits.mplot3d import Axes3D
         fig = figure()
         ax = Axes3D(fig)
         X = np.arange(-4, 4, 0.25)
         Y = np.arange(-4, 4, 0.25)
         X, Y = np.meshgrid(X, Y)
         R = np.sqrt(X**2 + Y**2)
         Z = np.sin(R)

         ax.plot_surface(X, Y, Z, rstride=1, cstride=1, cmap=cm.coolwarm)

Out[2]:  <mpl_toolkits.mplot3d.art3d.Poly3DCollection at 0x27416d0>
```

You can find more information about `matplotlib` from `http://matplotlib.org/`.

The second method for plotting in Wakari is through their custom plotting library `webplot` (still in development), which creates SVG graphics, currently just supporting line plots and scatter plots. In the following screenshot, we can observe an example of a scatter plot of random points using `webplot`:

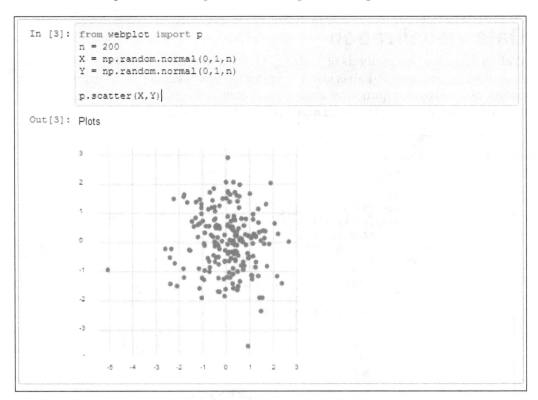

Introduction to image processing with PIL

The goal of this chapter is to present some of the preinstalled capabilities of Wakari. In this section, we will explore some of the basic functions of the **PIL** (**Python Image Library**) such as histogram, filters, operations, and transformations. We have already installed and used PIL in *Chapter 5, Similarity-based Image Retrieval*.

First, we will upload the images `412.jpg` (Dinosaur) and `826.jpg` (Land) to the path (see the arrow in the following screenshot). The images came from the **Caltech-256** images-dataset used in the *Chapter 5, Similarity-based Image Retrieval*.

Opening an image

The first thing we need to start working on is importing the `PIL` and `pylab` modules. Next, we will use the `open` method of the `Image` object. Finally, we will visualize the image with the `imshow` method of `pylab`. In the following screenshot, we may see the output of the code:

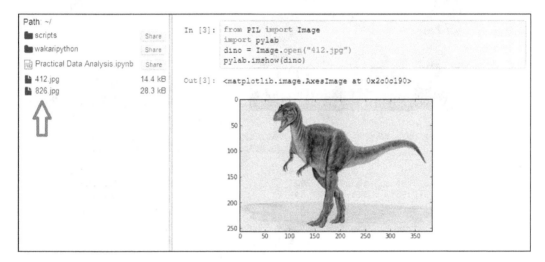

> You can find more information about PIL from
> `http://www.pythonware.com/products/pil/`.

Image histogram

A histogram is the distribution of the frequency of the intensity of each pixel. PIL provide us with a `histogram` method, which will get the frequency of each tone of color. As our images are in **RGB (Red, Green, Blue)**, we will get an array of 768 values (256 tones x 3 colors).

Often, we will need the histogram of a grayscale image because it will be easier to work with only 256 values of gray intensity instead of the full RGB color model. In PIL we just add the `L` parameter to the `histogram` method and the image will be treated as a grayscale image:

```
hist = land.histogram("L")
```

In the following screenshot, we will get the RGB histogram of the image (826.jpg) and we will plot the histogram using the hist method of pylab:

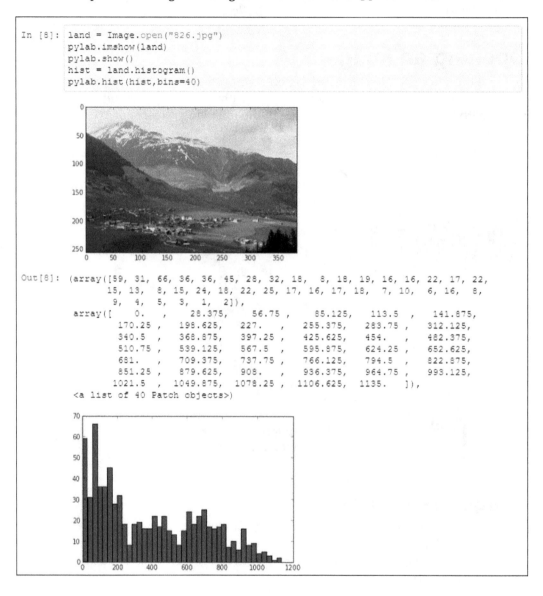

Filtering

The `filter` method will return a copy of the image filtered by the given filter. We will use the `ImageFilter` object, which currently supports the BLUR, CONTOUR, DETAIL, EDGE_ENHANCE, EDGE_ENHANCE_MORE, EMBOSS, FIND_EDGES, SMOOTH, SMOOTH_MORE, and SHARPEN filters. In this section, we will test some of the common filters and plot them with the `imshow` method of `pylab`. In the following screenshot, we observe the BLUR filter applied to the image of the dinosaur:

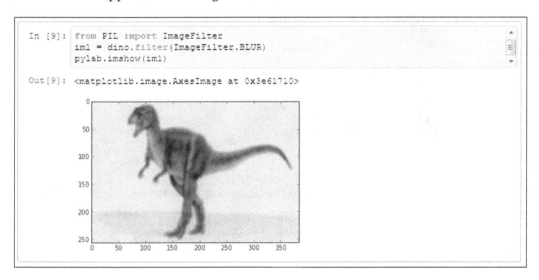

In the following screenshot, we observe the FIND_EDGES filter applied to the image of the dinosaur:

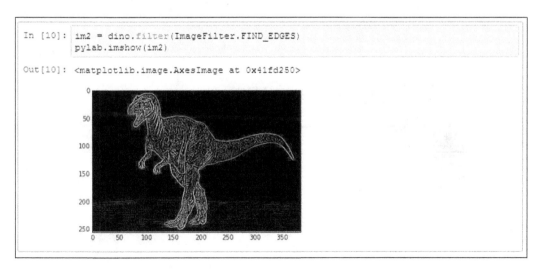

In the following screenshot, we observe the EDGES_ENHANCE_MORE filter applied to the image of the land:

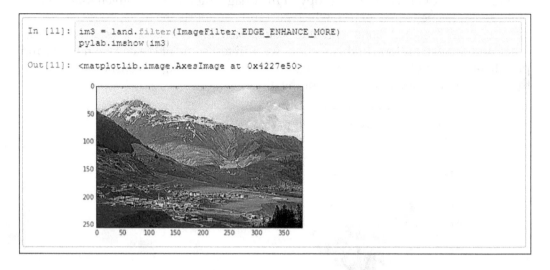

In the following screenshot, we observe the COUNTOUR filter applied to the image of the land:

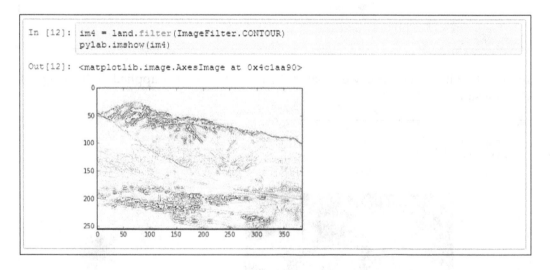

 For reference documentation of ImageFilter object, visit http://bit.ly/1fenKFq.

Operations

PIL include some of the most common image processing operations ready to be used with the `ImageOps` object.

In the following screenshot, we can see the dinosaur image using the `invert` operation, which inverts each pixel value (photographic negative). We will use the `invert` method from the `ImageOps` object included in the PIL library:

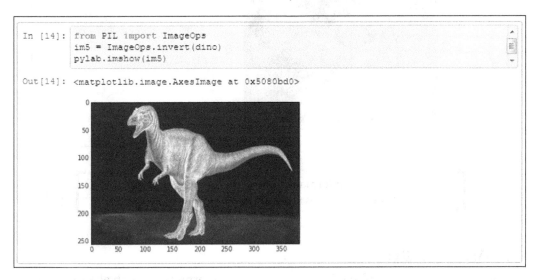

In the following screenshot, we can see the dinosaur image converted to grayscale:

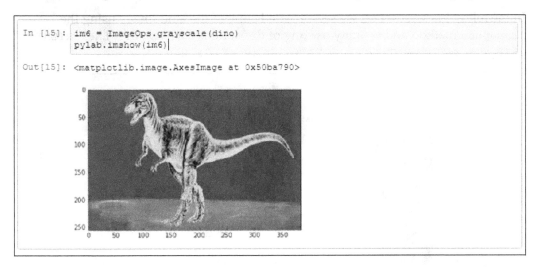

In the following screenshot, we can see the dinosaur image using a `solarize` method, which inverts all the pixel values above a given threshold:

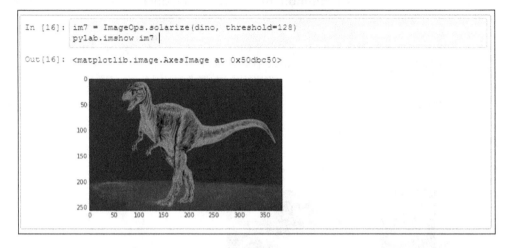

  For more information about the `ImageOps` object, visit `http://bit.ly/1741meW`.

Transformations

PIL provide us with several methods for image transformations such as `transform`, `transpose`, `crop`, and so on.

In the following screenshot, we see a rotated copy of the land image using the `transpose` method and we may use any of the options such as `FLIP_LEFT_RIGHT`, `FLIP_TOP_BOTTOM`, `ROTATE_90`, `ROTATE_180`, or `ROTATE_270`.

```
In [17]: im8 = land.transpose(Image.ROTATE_270)
         pylab.imshow(im8)

Out[17]: <matplotlib.image.AxesImage at 0x582f090>
```

In the following screenshot, we may see a rectangular region of the land image using the `crop` method, which receives a list with the pixel coordinates (left, upper, right, and lower). The `crop` method returns a copy of the rectangular region from the image:

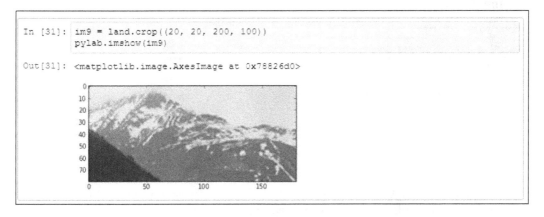

Getting started with Pandas

Pandas is a great library for data manipulation and analysis, written by Wes McKinney. Pandas provide us with optimized data structures such as **Series** and **DataFrame**, which are well suited for descriptive statistics, indexing, and aggregation. Pandas is already installed in the Anaconda distribution used in Wakari. In this section, we will present the basic operations with Pandas for time series and multivariate data. We may find more information about Pandas at `http://pandas.pydata.org/`.

Working with time series

Time series helps us to understand the change in a variable through time. Pandas include specific functionality in order to work with time series transparently. For this section, we need to upload the `Gold.csv` file used in *Chapter 7, Predicting Gold Prices*. The first five rows in the file will look as follows:

```
date,price
1/31/2003,367.5
2/28/2003,347.5
3/31/2003,334.9
4/30/2003,336.8
5/30/2003,361.4
. . .
```

We will load the `Gold.csv` file with the `read_csv` method (previously uploaded to the path of your account) and we will parse the dates just by activating the `parse_date` parameter (`parse_dates=True`). In the following screenshot, we can see that the result of the loading is a `DataFrame` object with a `DatetimeIndex` and a data column with the price:

```
In [1]: import pandas as pd
        ts = pd.read_csv('Gold.csv', index_col=0, parse_dates=True)
        ts

Out[1]: <class 'pandas.core.frame.DataFrame'>
        DatetimeIndex: 125 entries, 2003-01-31 00:00:00 to 2013-05-31 00:00:00
        Data columns (total 1 columns):
        price     125  non-null values
        dtypes: float64(1)
```

Next, we will plot the time series simply by calling the `plot` method of our `DataFrame`. In the following screenshot, we may see the gold prices from 2003 to 2013. The `plot` method of the `DataFrame` is a wrapper of `plt.plot` method of the `malplotlib` library:

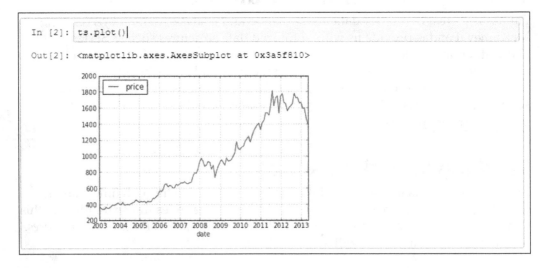

We can slice the time series simply by specifying a range. In case of the following screenshot, we just plot the records between 2006 and 2007 (`["2006":"2007"]`):

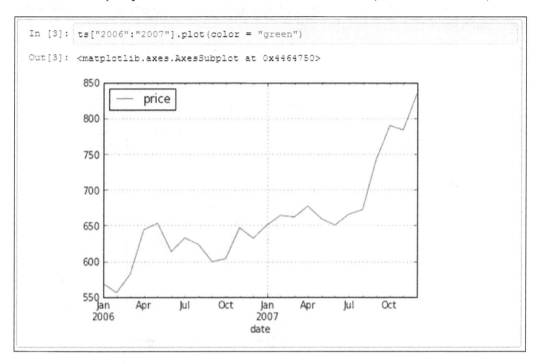

```
In [3]:  ts["2006":"2007"].plot(color = "green")

Out[3]:  <matplotlib.axes.AxesSubplot at 0x4464750>
```

We may also define a specific date `ts["2003/05/30"]` or a specific month `ts["2003/05"]`. Time series can be also sliced between two dates using the `truncate` method:

```
ts.truncate(after = "05/30/2003")
```

Pandas provide us with flexible resampling operations to perform frequency (monthly, yearly, weekly, daily, and so on) conversion. In the following screenshot, we will convert monthly data into annual data using the `resample` method. We will see a much smoother series in the plot:

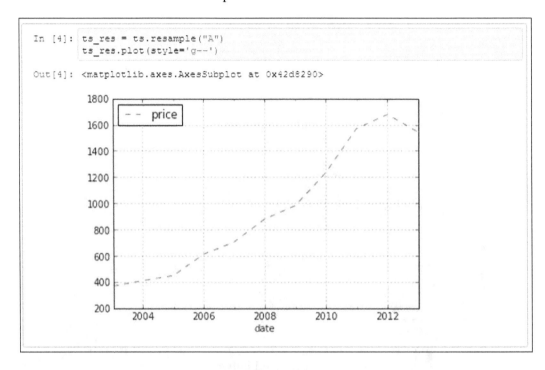

The `how` parameter of the `resample` method could be a custom function name or a NumPy array function that takes an array and produces aggregated data. For example, if we want only the `max` values, we will set the parameter as follows:

```
ts.resample("A", how=[np.max])
```

In the following screenshot, we will get three series; mean, max, and min. We will plot them in two different ways, the first one is with the subplots=True option, which will display three different figures, and the second one is the direct plot in which we will see three lines in the same figure:

```
In [5]:  ts_res = ts.resample("A", how=["mean", np.max, np.min])
         ts_res.plot(subplots=True)
         ts_res.plot()

Out[5]:  <matplotlib.axes.AxesSubplot at 0x4daf710>
```

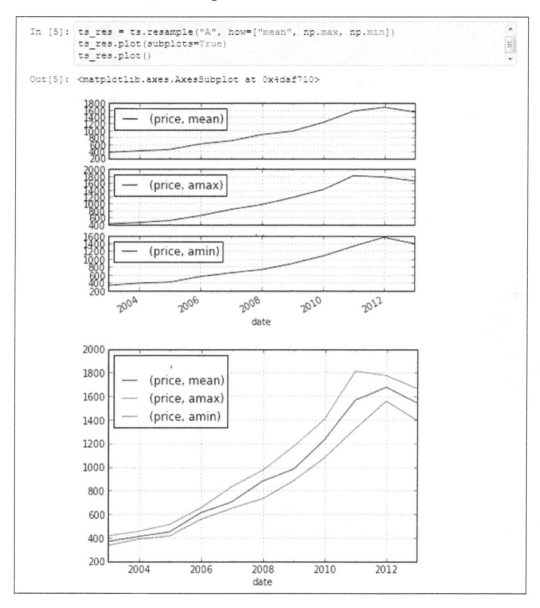

[For more on Pandas time series documentation visit `http://pandas.pydata.org/pandas-docs/dev/timeseries.html`.]

Working with multivariate dataset with DataFrame

In this section, we will perform some descriptive statistics with a multivariate dataset using a Pandas `DataFrame` object. In this section, we will use the `iris.csv` dataset; due to this we need to upload the file into the Wakari path before we start working on our IPython Notebook. The iris flower dataset is probably the most used dataset for classification with three categories (`setosa`, `versicolour`, `virginica`), four attributes (`SepalLength`, `SepalWidth`, `PetalLength`, `PetalWidth`), and 150 rows. We can download the iris dataset from UC Irvine Machine Learning Repository, available at `http://archive.ics.uci.edu/ml/datasets/Iris`.

The first five records in the `iris.csv` file will look as follows:

```
name,SepalLength,SepalWidth,PetalLength,PetalWidth
setosa,5.1,3.5,1.4,0.2
setosa,4.9,3,1.4,0.2
setosa,4.7,3.2,1.3,0.2
setosa,4.6,3.1,1.5,0.2
setosa,5,3.6,1.4,0.2
. . .
```

First, we need to load the `iris.csv` file into a `DataFrame` object using the `read_csv` method. Then, we will plot the dataset using **RadViz**, which is a radial visualization that can help us to visualize a multivariate data. The visualized attributes are presented as anchor points, equally split around the perimeter of the circle, and in the following screenshot we may see the `SepalLength`, `SepalWidth`, `PetalLength`, `PetalWidth` anchors. The dataset instances (rows) are shown as points inside the circle and this visualization can be used as a classification technique. In the following screenshot, we can see the plot of the iris dataset using `radviz` method:

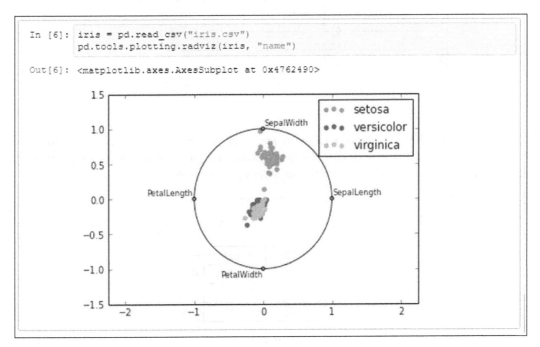

```
In [6]: iris = pd.read_csv("iris.csv")
        pd.tools.plotting.radviz(iris, "name")

Out[6]: <matplotlib.axes.AxesSubplot at 0x4762490>
```

Pandas provide us with the `head` method (see the following screenshot), which will get the first five records of our `DataFrame` and the `tail` method, which will get the last five records:

```
In [7]: iris.head()
Out[7]:
```

	name	SepalLength	SepalWidth	PetalLength	PetalWidth
0	setosa	5.1	3.5	1.4	0.2
1	setosa	4.9	3.0	1.4	0.2
2	setosa	4.7	3.2	1.3	0.2
3	setosa	4.6	3.1	1.5	0.2
4	setosa	5.0	3.6	1.4	0.2

We can get basic statistics from the `DataFrame` object with the `max`, `min`, and `mean` methods individually. But we can also get a summary of the `DataFrame` object using the `describe` method as shown in the following screenshot:

```
In [10]: iris.describe()
Out[10]:
```

	SepalLength	SepalWidth	PetalLength	PetalWidth
count	150.000000	150.000000	150.000000	150.000000
mean	5.843333	3.057333	3.758000	1.199333
std	0.828066	0.435866	1.765298	0.762238
min	4.300000	2.000000	1.000000	0.100000
25%	5.100000	2.800000	1.600000	0.300000
50%	5.800000	3.000000	4.350000	1.300000
75%	6.400000	3.300000	5.100000	1.800000
max	7.900000	4.400000	6.900000	2.500000

With a scatterplot, we see the correlation between two variables. However, when we have a multivariate dataset, the number of scatter plots increase. In these cases, we can use a scatterplot matrix in order to make easier to plot the correlations of a dataset.Pandas provide us with `scatter_matrix` method in `pandas.tools.plotting` (see the following screenshot):

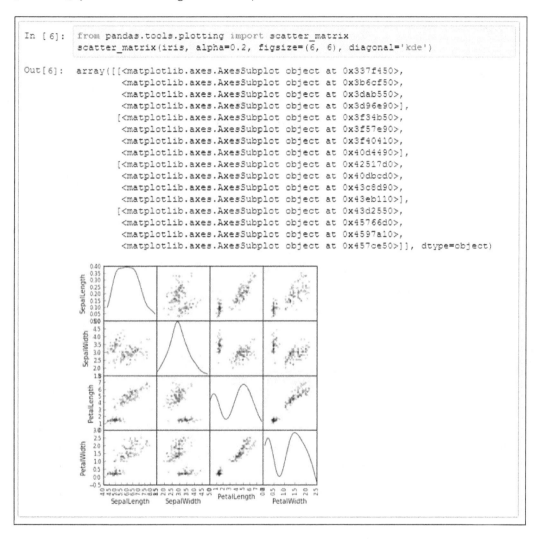

 You can find Pandas `DataFrame` documentation at
http://pandas.pydata.org/pandas-docs/dev/dsintro.html.

Grouping, aggregation, and correlation

Pandas provide us with syntactic sugar for grouping and aggregation of a
DataFrame object simply by applying the groupby method and selecting a column
for the grouping:

```
g = iris.groupby("name")
for name, group in g: print name
```

>>>**setosa**

>>>**versicolor**

>>>**virginica**

In the following screenshot, we can see the aggregated data using the sum, max and
min methods for the dataset grouped by name.

In [11]: `iris.groupby("name").sum()`

Out[11]:

name	SepalLength	SepalWidth	PetalLength	PetalWidth
setosa	250.3	171.4	73.1	12.3
versicolor	296.8	138.5	213.0	66.3
virginica	329.4	148.7	277.6	101.3

In [12]: `iris.groupby("name").max()`

Out[12]:

name	SepalLength	SepalWidth	PetalLength	PetalWidth
setosa	5.8	4.4	1.9	0.6
versicolor	7.0	3.4	5.1	1.8
virginica	7.9	3.8	6.9	2.5

In [13]: `iris.groupby("name").min()`

Out[13]:

name	SepalLength	SepalWidth	PetalLength	PetalWidth
setosa	4.3	2.3	1.0	0.1
versicolor	4.9	2.0	3.0	1.0
virginica	4.9	2.2	4.5	1.4

We may also call the `describe` method for the grouped data as shown in the following screenshot. In this case, we will get the aggregated data for each group:

In [14]: `iris.groupby("name").describe()`

Out[14]:

name		SepalLength	SepalWidth	PetalLength	PetalWidth
setosa	count	50.000000	50.000000	50.000000	50.000000
	mean	5.006000	3.428000	1.462000	0.246000
	std	0.352490	0.379064	0.173664	0.105386
	min	4.300000	2.300000	1.000000	0.100000
	25%	4.800000	3.200000	1.400000	0.200000
	50%	5.000000	3.400000	1.500000	0.200000
	75%	5.200000	3.675000	1.575000	0.300000
	max	5.800000	4.400000	1.900000	0.600000
versicolor	count	50.000000	50.000000	50.000000	50.000000
	mean	5.936000	2.770000	4.260000	1.326000
	std	0.516171	0.313798	0.469911	0.197753
	min	4.900000	2.000000	3.000000	1.000000
	25%	5.600000	2.525000	4.000000	1.200000
	50%	5.900000	2.800000	4.350000	1.300000
	75%	6.300000	3.000000	4.600000	1.500000
	max	7.000000	3.400000	5.100000	1.800000
virginica	count	50.000000	50.000000	50.000000	50.000000
	mean	6.588000	2.974000	5.552000	2.026000
	std	0.635880	0.322497	0.551895	0.274650
	min	4.900000	2.200000	4.500000	1.400000
	25%	6.225000	2.800000	5.100000	1.800000
	50%	6.500000	3.000000	5.550000	2.000000
	75%	6.900000	3.175000	5.875000	2.300000
	max	7.900000	3.800000	6.900000	2.500000

We can also group by multiple attributes, as shown in the following code:

```
for name, group in iris.groupby(["name", "SepalLength"]):
    print name
    print group
```

The result groups will look as follows:

```
('setosa', 4.3)
        name  SepalLength  SepalWidth  PetalLength  PetalWidth
13    setosa          4.3           3          1.1         0.1
('setosa', 4.4)
        name  SepalLength  SepalWidth  PetalLength  PetalWidth
8     setosa          4.4         2.9          1.4         0.2
38    setosa          4.4         3.0          1.3         0.2
42    setosa          4.4         3.2          1.3         0.2

. . .
```

 You find the Pandas groupby method documentation at
http://pandas.pydata.org/pandas-docs/dev/groupby.html.

Pandas DataFrame provides us with a correlation function (corr) and implements three different correlation coefficient methods; pearson (default), kendall, and spearman using the method parameter:

```
iris.corr(method='spearman')
```

In this case, we will get the correlation between two attributes (see In[15] in the following screenshot) and the correlation of all the attributes (see In[16] in the following screenshot):

```
In [15]:  iris["SepalLength"].corr(iris["PetalLength"])

Out[15]:  0.87175377588658287

In [16]:  iris.corr()

Out[16]:
```

	SepalLength	SepalWidth	PetalLength	PetalWidth
SepalLength	1.000000	-0.117570	0.871754	0.817941
SepalWidth	-0.117570	1.000000	-0.428440	-0.366126
PetalLength	0.871754	-0.428440	1.000000	0.962865
PetalWidth	0.817941	-0.366126	0.962865	1.000000

Multiprocessing with IPython

In data analysis, we often perform processing tasks which are computationally expensive. In these cases we will need multiprocessing tools that enable us to improve the performance. Multiprocessing in IPython is a big enough topic to have its own chapter. In this section, we only show how we can run a map function into parallel processes with the Pool object in Wakari.

Pool

The Pool class is the easiest way to run a parallel process into a Wakari IPython Notebook. In this case, we will create a function that will be applied to each element on a numpy array by using the map_async method, which is a variant of the map method that delivers the result asynchronously.

In the following screenshot, we can see the result of the map_async function of the Pool object. With the get method, we will get the result when it arrives:

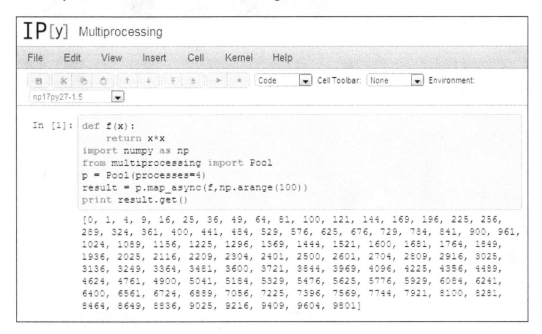

 You can find the multiprocessing module documentation at
http://docs.python.org/2/library/multiprocessing.html.

Sharing your Notebook

One of the most amazing features of Wakari is that we can share our notebooks with other Wakari users and they can import it into their accounts. This feature makes Wakari an excellent choice for teaching a workshop or for a presentation.

The data

When our IPython Notebook is ready, we can share it with other Wakari users just by clicking on the **Share** button, next to the name of our notebook in the resources tab.

In the following screenshot we can see the **Sharing** window, where we may change the name and add a description to our notebook. For paid accounts, we can also include a password to keep our notebook private.

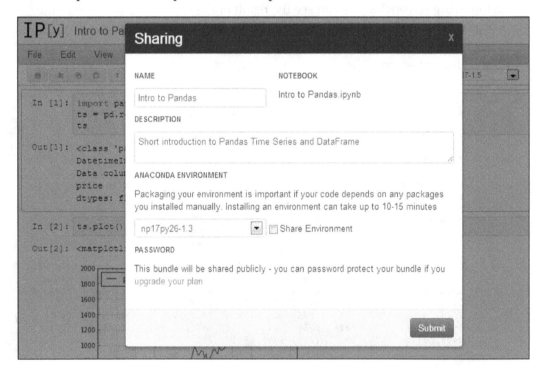

Once we are ready, we will click on the **Submit** button. We will see in the **Sharing Status** window that the process is complete and we can click on **Link to the bundle** to see our notebook shared (see the following screenshot):

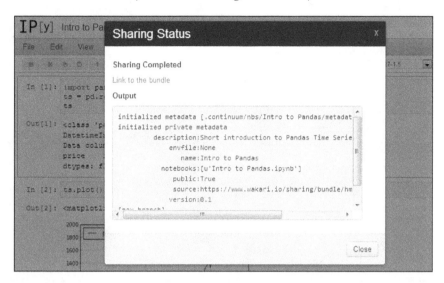

After clicking on **Link to the bundle**, we will see our IPython Notebook **Intro to Pandas** as a read-only file. If we click on the **Run/Edit this Notebook** button, we will create a copy of the notebook in our Wakari environment that we can upload freely:

In the following screenshot, we can see our **Shared Bundles** by navigating to
Account name | Settings | Sharing, there we can get the link or delete our
shared notebook:

Wakari also provides us with a gallery in which we can find good tutorials
as notebooks that we can copy and modify. You can find the gallery at
`https://www.wakari.io/gallery`.

All the codes and notebooks of this chapter may be found in
the author's GitHub repository at `https://github.com/
hmcuesta/PDA_Book/tree/master/Chapter14`.

Summary

In this chapter, we have explored an interesting tool for online data analysis with Python. Wakari provided us with a scientific environments ready to use, which is a great tool for teaching and sharing code. In this chapter, we provided a small introduction to image processing and to the Pandas library. In Pandas we learned how to work with time series and multivariate dataset. Finally, we learned how to share our IPython Notebooks with others Wakari users.

Wakari is highly recommended for all the Python community because it provides a robust Anaconda environment out of the box and supports all the major Python libraries.

Setting Up the Infrastructure

This chapter includes instructions for installing and configuring software packages that support all the projects in this book.

In this chapter, we will cover the installations of:

- Python 3
- IDLE
- Numpy
- SciPy
- mlpy
- OpenRefine
- MongoDB
- UMongo
- Gephi

Installing and running Python 3

Python is a general-purpose programming language whose design philosophy emphasizes batteries included, which provides clear and logical programs on small and large scale.

The latest versions of Ubuntu and Fedora come with Python 2.7 out of the box. In this book, we will use Python 3.2 for the code examples and projects. Python comes with a large set of standard libraries that support many common programming tasks such as collections, connecting to web servers, high-performance scientific computing, searching text with regular expressions, reading and modifying files.

We will make use of several Python libraries such as `numpy`, `scipy`, `mlpy`, `nose`, `pymongo`. In this chapter, we will see how to install and set up all these libraries. We can find more information on the Python's official website, `http://python.org/`.

Installing and running Python 3.2 on Ubuntu

To install python, simply open a command prompt and run the following command:

```
$ sudo apt-get install python3
```

To check whether everything is installed correctly, just execute the following command:

```
$ python3
```

```
●●● packt@packt-PC: ~
packt@packt-PC:~$ python3
Python 3.2.3 (default, Sep 30 2012, 16:43:30)
[GCC 4.7.2] on linux2
Type "help", "copyright", "credits" or "license" for more information.
>>>
```

Installing and running IDLE on Ubuntu

To install IDLE, just open a command prompt and run the following command:

```
$ sudo apt-get install idle3
```

To check whether everything is installed correctly, just execute the following command:

```
$ idle3
```

```
●●● Python Shell
File  Edit  Shell  Debug  Options  Windows  Help
Python 3.2.3 (default, Sep 30 2012, 16:43:30)
[GCC 4.7.2] on linux2
Type "copyright", "credits" or "license()" for more information.
>>>
>>> |
                                                        Ln: 5 Col: 4
```

Installing and running Python 3.2 on Windows

First, download Python 3.2 from the official website, `http://www.python.org/download/releases/3.2.3/`.

The Windows version is provided as an MSI package. To install it manually, just double click the `/python-3.2.3.msi` file:

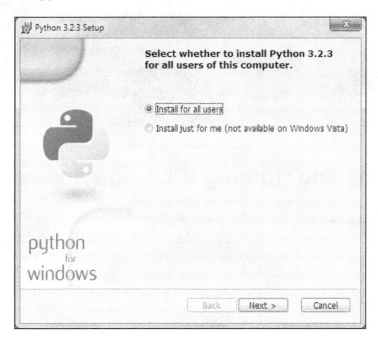

By design, Python installs to a directory with the version number embedded. In this case, Python version is 3.2 and will install at `C:\Python32\`, so that you can have multiple versions of Python on the same system without any conflicts.

Python does not automatically modify the `PATH` environment variable, so you will need to do it manually. Right-click on **My Computer**, select **Properties**, **Advance System Settings**, and click on the **Environment Variables** button.

Now edit the `PATH` system-variable and add `;C:\Python32\;C:\Python32\Scripts\` to its end.

To check whether everything is installed correctly, just execute the following command in the Windows terminal:

```
>> python
```

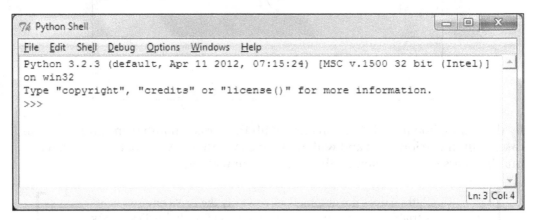

Installing and running IDLE on Windows

IDLE is already installed with Python MSI installation, to run it just navigate to **Start | All Programs | Python 3.2 | IDLE (Python GUI)**:

The `easy_install` command makes it easy to fetch and install Python libraries and their dependencies. The most crucial third-party Python software of all is **Distribute**, which extends the packaging and installation facilities provided by `distutils` in the standard library.

To obtain the latest version of Distribute for Windows, run the Python script available at `http://www.lfd.uci.edu/~gohlke/pythonlibs/#distribute`. Download and execute `distribute-0.6.35.win32-py3.2.exe`. Now `easy_install` gets installed into `c:\Python32\Scripts`.

Installing and running NumPy

According to the official website `http://www.numpy.org/`, NumPy is the fundamental package for scientific computing with Python. It contains amongst other things:

- A powerful N-dimensional array object
- Sophisticated (broadcasting) functions
- Tools for integrating C/C++ and Fortran code
- Useful linear algebra, Fourier transform, and random number capabilities

Besides its obvious scientific uses, NumPy can also be used as an efficient multi-dimensional container of the generic data. Arbitrary datatypes can be defined. This allows NumPy to seamlessly and speedily integrate with a wide variety of datasets.

Installing and running NumPy on Ubuntu

To install numpy, simply open a command prompt and run.

```
$ sudo apt-get install python3-numpy
```

To check whether everything is installed correctly, just execute the Python Shell shown as follows:

```
$ idle3
```

Then execute the following commands:

```
>>> import numpy
>>> numpy.test()
```

> We need to use the `nose` library (that extends the test loading and running features of unit test) when using `numpy.test()`.
>
> In order to install it, we just need to open a command line and run the following command:
>
> **$ sudo apt-get install python3-nose**
>
> Or you can also execute:
>
> **$ pip install nose**
>
> For more information about the nose library, visit `https://pypi.python.org/pypi/nose/1.1.2`.

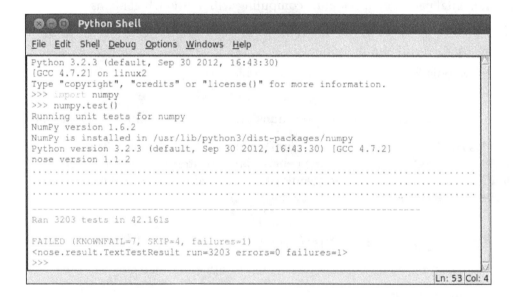

```
Python 3.2.3 (default, Sep 30 2012, 16:43:30)
[GCC 4.7.2] on linux2
Type "copyright", "credits" or "license()" for more information.
>>> import numpy
>>> numpy.test()
Running unit tests for numpy
NumPy version 1.6.2
NumPy is installed in /usr/lib/python3/dist-packages/numpy
Python version 3.2.3 (default, Sep 30 2012, 16:43:30) [GCC 4.7.2]
nose version 1.1.2
............................................................................
............................................................................
............................................................................
----------------------------------------------------------------------
Ran 3203 tests in 42.161s

FAILED (KNOWNFAIL=7, SKIP=4, failures=1)
<nose.result.TextTestResult run=3203 errors=0 failures=1>
>>>
```

Installing and running NumPy on Windows

First, download the NumPy 1.7 from the official website `http://sourceforge.net/projects/numpy/files/NumPy/1.7.0/`.

The Windows version is provided as an `.exe` package. To install it manually, just double click on the `/numpy-1.7.0-win32-superpack-python3.2.exe` file.

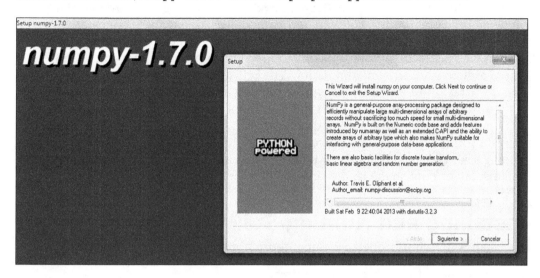

To check whether everything is installed correctly, just navigate to **Start | All Programs | Python 3.2 | IDLE (Python GUI)**.

Then execute the following commands:

```
>>> import numpy
>>> numpy.test()
```

We need to use the `nose` library (that extends the test loading and running features of unit test) when using `numpy.test()`.

In order to install it, you just need to open a Windows command line (CMD) and run the following command:
`C:\> pip install nose`

For more information about `nose`, visit `https://pypi.python.org/pypi/nose/1.1.2`.

Installing and running SciPy

According to the official website http://www.scipy.org/, SciPy (pronounced as Sigh Pie) is an open-source software for mathematics, science, and engineering. It is also the name of a very popular conference on scientific programming with Python. The SciPy library depends on NumPy, which provides convenient and fast N-dimensional array manipulation. The SciPy library is built to work with NumPy arrays, and provides many user-friendly and efficient numerical routines such as routines for numerical integration and optimization. Together, they run on all popular operating systems, are quick to install, and are free of charge. NumPy and SciPy are easy to use, and powerful enough to be used by some of the world's leading scientists and engineers. If you need to manipulate numbers on a computer, and display or publish the results, give SciPy a try!

Installing and running SciPy on Ubuntu

To install SciPy, simply open a command prompt and run the following command:

```
$ sudo apt-get install python3-scipy
```

To check whether everything is installed correctly, just execute the Python Shell as follows:

```
$ idle3
```

Then execute the following commands:

```
>>> import scipy
>>> scipy.test()
```

```
Python Shell
File  Edit  Shell  Debug  Options  Windows  Help
Python 3.2.3 (default, Sep 30 2012, 16:43:30)
[GCC 4.7.2] on linux2
Type "copyright", "credits" or "license()" for more information.
>>> import scipy
>>> scipy.test()
Running unit tests for scipy
NumPy version 1.6.2
NumPy is installed in /usr/lib/python3/dist-packages/numpy
SciPy version 0.10.1
SciPy is installed in /usr/lib/python3/dist-packages/scipy
Python version 3.2.3 (default, Sep 30 2012, 16:43:30) [GCC 4.7.2]
nose version 1.1.2
.................................................................K..K..
........F.................................
----------------------------------------------------------------------
Ran 4138 tests in 84.278s

FAILED (KNOWNFAIL=11, SKIP=26, failures=1)
<nose.result.TextTestResult run=4138 errors=0 failures=1>
>>>
                                                              Ln: 22 Col: 36
```

Installing and running SciPy on Windows

First, download the SciPy 0.12 from the official website, `http://sourceforge.net/projects/scipy/files/scipy/0.12.0b1/`.

The Windows version is provided as an `.exe` package. To install it manually, just double click the `/scipy-0.12.0b1-win32-superpack-python3.2.exe/` file.

To check whether everything is installed correctly, just navigate to **Start | All Programs | Python 3.2 | IDLE (Python GUI)**.

Then execute the following commands:

```
>>> import scipy
>>> scipy.test()
```

Installing and running mlpy

According with the official website `http://mlpy.sourceforge.net/`, mlpy provides a wide range of state-of-the-art machine learning methods for supervised and unsupervised problems. It is aimed at finding a reasonable compromise among modularity, maintainability, reproducibility, usability, and efficiency. Mlpy is multiplatform, it works with Python 2 and 3, and it is open source. Mlpy is distributed under the GNU General Public License Version 3.

We need the following requirements:

- GCC
- Numpy 1.7
- SciPy 0.12
- GSL 1.11

Installing and running mlpy on Ubuntu

First, we need to download the latest version for Linux from `http://sourceforge.net/projects/mlpy/files/mlpy%203.5.0/`.

Unzip and run the following command from the terminal:

```
$ sudo python3 setup.py install
```

 The installation requires GSL 1.11 or greater. We can install the library from Ubuntu Software Center. We just need to look and install the **GNU Scientific Library (GSL) development package**.

To check whether everything is installed correctly, just open a Python shell and execute:

```
>>> import mlpy
```

Installing and running mlpy on Windows

First, we need to download the latest version for Windows from `http://sourceforge.net/projects/mlpy/files/mlpy%203.5.0/`.

Then execute the `mlpy-3.5.0.win32-py3.2.exe` file and follow the wizard as shown in the following screenshot:

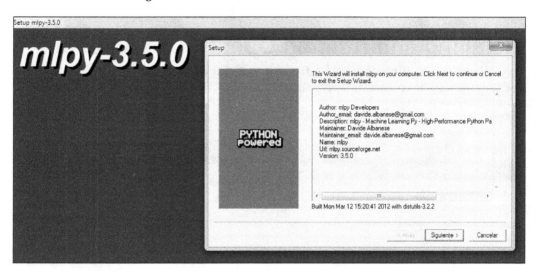

 The GSL library is precompiled (by **Visual Studio Express 2008**) and included in `mlpy`.

To check whether everything is installed correctly, just open a Python shell and execute the following command:

```
>>> import mlpy
```

Installing and running OpenRefine

According to the official website `http://openrefine.org/`, OpenRefine (ex-Google Refine) is a powerful tool for working with messy data, cleaning it, transforming it from one format into another, extending it with web services, and linking it to databases such as Freebase.

See *Chapter 3*, *Data Visualization*, for detailed instructions about how to clean, normalize, and export data using OpenRefine.

In order to run OpenRefine on Windows or Linux, you need to have installed either Java Standard Edition or OpenJDK on the computer.

You can download the latest version of JSE from the official website, `http://www.oracle.com/technetwork/java/javase/downloads/`.

Installing and running OpenRefine on Linux

First, download the OpenRefine 2.5 from the official website, `http://google-refine.googlecode.com/files/google-refine-2.5-r2407.tar.gz`.

Extract, open a terminal in the directory, then type `./refine` to start.

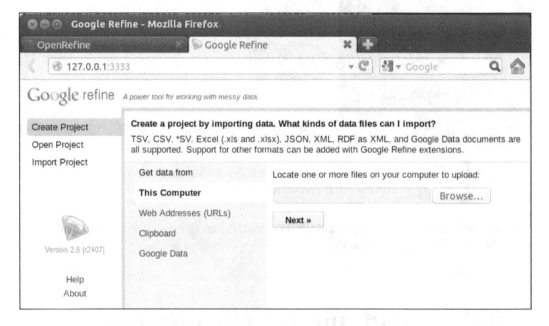

Installing and running OpenRefine on Windows

First, download the OpenRefine 2.5 from the official website, `http://google-refine.googlecode.com/files/google-refine-2.5-r2407.zip`.

Unzip, and double-click on `google-refine.exe`.

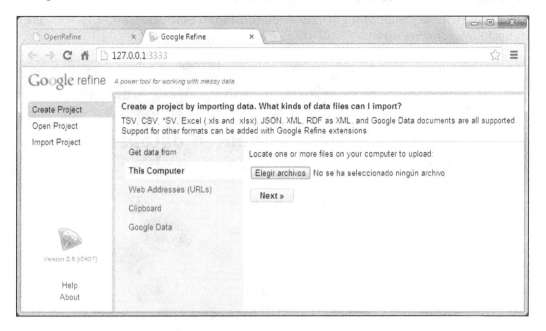

Installing and running MongoDB

According to the official website `http://www.mongodb.org/`, MongoDB (from humongous) is an open source document database, and the leading NoSQL database. Written in C++, MongoDB features:

- **Document-oriented storage**: JSON-style documents with dynamic schemas that offer simplicity and power

- **Full index support**: Index on any attribute, just like you're used to

- **Replication and high availability**: Mirror across LANs and WANs for scale and peace of mind

- **Auto-sharding**: Scale horizontally without compromising functionality

- **Querying**: Rich document-based queries

- **Fast in-place updates**: Atomic modifiers for contention-free performance

- **Map/Reduce**: Flexible aggregation and data processing

- **GridFS**: Store files of any size without complicating your stack

- **Commercial support**: Enterprise class support, training, and consulting available

Installing and running MongoDB on Ubuntu

The easiest way to install MongoDB is through Ubuntu Software Center, as showed in the following screenshot:

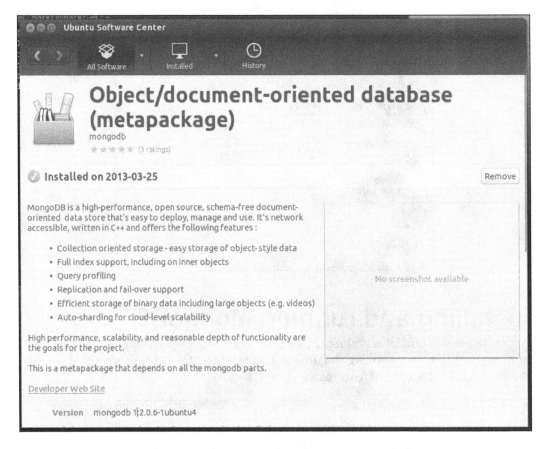

Finally, just open a terminal and execute mongo, as shown in the following screenshot:

```
$ mongo
```

```
packt@packt-PC: ~
packt@packt-PC:~$ mongo
MongoDB shell version: 2.0.6
connecting to: test
>
```

To check whether everything is installed correctly, just execute the Mongo shell as shown in the following screenshot. Insert a record in the test collection and retrieve that record:

```
> db.test.save( { a: 1 } )
> db.test.save( { a: 100 } )
> db.test.find()
```

```
packt@packt-PC: ~
packt@packt-PC:~$ mongo
MongoDB shell version: 2.0.6
connecting to: test
> db.test.save({a:100})
> db.test.find()
{ "_id" : ObjectId("5150bc5a541a01b67ab6d55e"), "a" : 1 }
{ "_id" : ObjectId("5150bd57119eaaf68edaebb2"), "a" : 100 }
>
```

Installing and running MongoDB on Windows

Download the latest production release of MongoDB from the official website, http://www.mongodb.org/downloads.

There are two builds of MongoDB for Windows:

- MongoDB for Windows 64-bit runs on any 64-bit version of Windows newer than Windows XP, including Windows Server 2008 R2 and Windows 7 64-bit.

- MongoDB for Windows 32-bit runs on any 32-bit version of Windows newer than Windows XP. 32-bit versions of MongoDB are only used in testing and development systems (is limited to less of 2GB for storage capacity).

Unzip in a folder such as `c:\mongodb\`.

MongoDB requires a data folder to store its files:

`C:\data\db`

Then to start MongoDB, we need to execute `mongod.exe` from the command prompt (`c:\mongodb\bin\mongod.exe`) as shown in the following screenshot:

```
C:\mongodb\bin\mongod.exe

C:\mongodb\bin\mongod.exe --help for help and startup options
Mon Mar 25 12:45:14.275
Mon Mar 25 12:45:14.276 warning: 32-bit servers don't have journaling enabled by
 default. Please use --journal if you want durability.
Mon Mar 25 12:45:14.276
Mon Mar 25 12:45:14.292 [initandlisten] MongoDB starting : pid=2416 port=27017 d
bpath=\data\db\ 32-bit host=Hadoop-PC
Mon Mar 25 12:45:14.293 [initandlisten]
Mon Mar 25 12:45:14.293 [initandlisten] ** NOTE: This is a 32 bit MongoDB binary
.
Mon Mar 25 12:45:14.293 [initandlisten] **       32 bit builds are limited to le
ss than 2GB of data (or less with --journal).
Mon Mar 25 12:45:14.293 [initandlisten] **       Note that journaling defaults t
o off for 32 bit and is currently off.
Mon Mar 25 12:45:14.293 [initandlisten] **       See http://dochub.mongodb.org/c
ore/32bit
Mon Mar 25 12:45:14.293 [initandlisten]
Mon Mar 25 12:45:14.293 [initandlisten] db version v2.4.1
Mon Mar 25 12:45:14.294 [initandlisten] git version: 1560959e9ce11a693be8b4d0d16
0d633eee75110
Mon Mar 25 12:45:14.294 [initandlisten] build info: windows sys.getwindowsversio
n(major=6, minor=0, build=6002, platform=2, service_pack='Service Pack 2') BOOST
_LIB_VERSION=1_49
Mon Mar 25 12:45:14.294 [initandlisten] allocator: system
Mon Mar 25 12:45:14.294 [initandlisten] options: {}
```

You can specify an alternate path for `c:\data\db`, with the `dbpath` setting for `mongod.exe`, as in the following example:

`C:\mongodb\bin\mongod.exe --dbpath c:\mongodb\data\`

You can get the full list of command-line options by running `mongod` with the `--help` option:

`C:\mongodb\bin\mongod.exe --help`

Finally, just execute mongo.exe and the Mongo browser shell is ready to use, as shown in the following screenshot:

```
C:\mongodb\bin\mongo.exe
```

```
Administrador: C:\Windows\System32\cmd.exe - mongo

C:\mongodb\bin>mongo
MongoDB shell version: 2.4.1
connecting to: test
Server has startup warnings:
Mon Mar 25 12:37:32.607 [initandlisten]
Mon Mar 25 12:37:32.608 [initandlisten] ** NOTE: This is a 32 bit MongoDB binary
Mon Mar 25 12:37:32.608 [initandlisten] **       32 bit builds are limited to le
ss than 2GB of data (or less with --journal).
Mon Mar 25 12:37:32.608 [initandlisten] **       Note that journaling defaults t
o off for 32 bit and is currently off.
Mon Mar 25 12:37:32.608 [initandlisten] **       See http://dochub.mongodb.org/c
ore/32bit
Mon Mar 25 12:37:32.608 [initandlisten]
>
```

 MongoDB is running on the localhost interface and port 27017 by default. If you want to change the port, you need to use the -port option of the mongod command.

To check whether everything is installed correctly, just run the Mongo shell as shown in the following screenshot. Insert a record in the test collection and retrieve that record:

```
> db.test.save( { a: 1 } )
> db.test.save( { a: 100 } )
> db.test.find()
```

```
> db.test.save(( a:100 ))
> db.test.find()
{ "_id" : ObjectId("51509a6dcceb8b15b98c5cc9"), "a" : 1 }
{ "_id" : ObjectId("51509b8d5ab6db3a5c3781c1"), "a" : 100 }
>
```

Connecting Python with MongoDB

The most popular module for working with MongoDB from Python is pymongo, it can be easily installed in Linux using pip, as shown in the following command:

```
$ pip install pymongo
```

> You may have installed multiple versions of Python. In that case, you may want to use virtualenv of Python3, and then install packages after activating virtualenv.
>
> Installing python-virtualenv:
> ```
> $ sudo apt-get install python-virtualenv
> ```
>
> Setting up the virtualenv:
> ```
> $ virtualenv -p /usr/bin/python3 py3env
> $ source py3env/bin/activate
> ```
>
> Installing packages for Python 3
> ```
> $ pip install "package-name"
> ```

In Windows, we can install pymongo using easy_install, opening a command prompt, and executing the following command:

```
C:/> easy-install pymongo
```

To check whether everything is installed correctly, just execute the Python shell as shown in the following code. Insert a record in the test_rows collection and retrieve that record:

```
>>> from pymongo import MongoClient
>>> con = MongoClient()
>>> db = con.test
>>> test_row = {'a':'200'}
>>> test_rows = db.rows
>>> test_rows.insert(test_row)
>>> result = test_rows.find()
>>> for x in result: print(x)
...
{'a':'200', 'id': ObjectId('5150c46b042a1824a78468b5')}
```

Installing and running UMongo

According to the official website `http://httpd.apache.org/`, UMongo is a GUI app that can browse and administer a MongoDB cluster. It is available for Linux, Windows, and Mac OSX.

Features of UMongo include:

- Connecting to a single server, a replica set, or a MongoS instance
- DB ops: create, drop, and authenticate, command, eval
- Collection ops: create, rename, drop, find, insert, save
- Document ops: update, duplicate, remove
- Index ops: create, drop
- Shard ops: enable sharding, add shard, shard collection
- GUI Document builder
- Import/export data from the database to local files in JSON, BSON, CSV format
- Support for query options and write concerns (`getLastError`)
- Display of numerous stats (server status, db stats, replication info, and so on)
- Mongo tree refreshes to have a real-time view of cluster (servers up/down, durability, and so on)
- All operations are executed in background to keep UI responsive
- Background threads can repeat commands automatically
- GUI is identical on all OS
- A login Screen
- User control management
- MySQL tables management (for categories, and combo-box values)
- Content management control
- Client e-mail module

 In order to run UMongo on Windows or Linux, you need to have Java Standard Edition installed on the computer.

You can download the latest version of JSE from the official website, `http://www.oracle.com/technetwork/java/javase/downloads/`.

See *Chapter 13, Working with MapReduce,* and *Chapter 14, Online Data Analysis with IPython and Wakari,* for detailed examples of Umongo.

Installing and running Umongo on Ubuntu

First, download the latest version of Umongo from the official website, `http://edgytech.com/wp-content/uploads/umongo-linux-all_1-2-1.zip`.

Extract the files, open the extracted folder, and double-click on `launch-umongo.sh`.

To check whether everything is installed correctly, we need to connect Umongo (**File/Connect**) with our mongo service, as shown in the following screenshot:

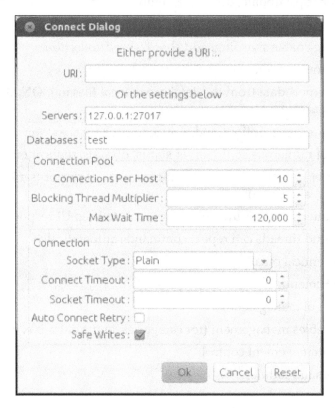

We need to input the server, port and database name. In the following screenshot, in the left we can find our databases and collections. With left-click over any collection, we can use the `find` command and the result will be set in the right tab:

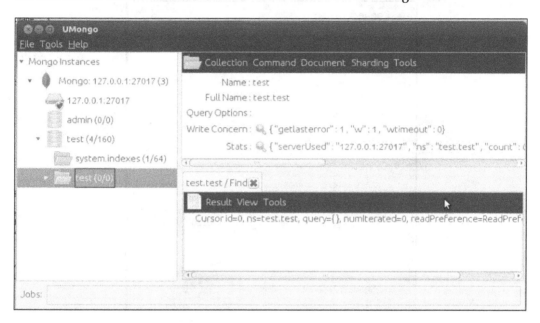

Installing and running Umongo on Windows

First, download the latest version of Umongo from the official website,
`http://edgytech.com/wp-content/uploads/umongo-windows-all_1-2-1.zip`.

Extract the files, open the extracted folder, and double-click on `umongo.exe`.

To check that everything is installed correctly, we need to connect Umongo (**File/Connect**) with our mongo service, as shown in the following screenshot:

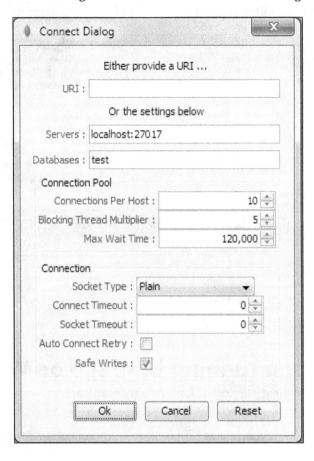

We need to provide with the server, port, and database name. In the following screenshot, in the left we can find our databases and collections. With left-click over any collection we can use the `find` command and the result will be set in the right tab:

Installing and running Gephi

According to the official website `https://gephi.org/`, Gephi is an interactive visualization and exploration platform for all kinds of networks and complex systems, dynamic and hierarchical graphs.

See *Chapter 10*, *Working with Social Graph*, for detailed instructions about how to use Gephi to visualize graphs.

Installing and running Gephi on Linux

First, download the Gephi 0.8.2 from the official website, `https://launchpad.net/gephi/0.8/0.8.2beta/+download/gephi-0.8.2-beta.tar.gz`.

Extract, open a terminal in the directory, and then type `./bin/gephi script file` to start.

Installing and running Gephi on Windows

First, download the Gephi 0.8.2 from the official website, `https://launchpad.net/` `gephi/0.8/0.8.2beta/+download/gephi-0.8.2-beta.setup.exe`.

Next, we need to execute the `setup.exe` file (see the following screenshot) and follow the wizard.

To check that everything is installed correctly, just navigate to **Start** | **All Programs** | **Gephi** | **Gephi 0.8.2**.

Index

e-mail validation 35
epidemic models
  about 156
  SIR model 156, 157
  SIRS model 159, 160
epidemiology 154
epidemiology triangle
  about 155
  Agent 155
  Environment 155
  Host 155
  Time 155
ETL 36
Euclidian distance 94
Excel files 28
explain method 232
Exploratory data analysis. *See* EDA
exploratory data analysis (EDA) 51, 141

F

Facebook graph
  acquiring 177
  acquiring, Netvizz used 178
Financial time series analysis. *See* FTSA
findall() function 36
find method 230
findOne method 230, 231
Flat window 124
followers, Twitter 201
followers, Twython
  working with 211
format function 72
formats, text files
  CSV 28
  JSON 28
  TSV 28
  XML 28
FTSA 105, 106
functionalities, MongoDB
  ad hoc queries 226
  aggregation 226
  load balancing 226
  Map-Reduce 226
  replication 226

G

GDF file
  transforming, to JSON format 190, 191
g element 62
genfromtxt function 39, 124
Gephi
  about 181, 323
  installing, on Linux 323
  installing, on Windows 324
  running, on Linux 323
  running, on Windows 324
  URL 181, 323
  used, for representing graphs 181, 182
GitHub repository
  URL 55
Global stochastic contact model 162
gold prices time series
  smoothing 129
Google Flu Trends data
  URL 155
Google Flu Trends (GFT) 154
Google Refine Expression Language. *See* GREL
graph
  about 175
  D3.js visualizations, creating 192-195
  directed graph 176
  representing, Gephi used 181, 182
  structure 175
  undirected graph 176
  uses 175
graph analytics
  about 175
  categories 176
  pattern-matching algorithms 176
  structural algorithms 176
  traversal algorithms 176
graph visualization
  D3.js used 192
GREL 47
groupby method 292
group() function 35
group function, MongoDB
  about 238
  using 238-241

grouping, MapReduce
  performing 259-261
grouping, Pandas 292-294

H

Hamming window 124
Hanning window 124
Hilary Mason. research-quality datasets
  URL 27
histogram 277
Historical Exchange Rates log
  URL 70
historical gold prices
  using 126
HTML
  about 53
  URL 53
HyperText Markup Language. *See* HTML

I

IDLE
  installing, on Ubuntu 302
  installing, on Windows 304
  running, on Ubuntu 302
  running, on Windows 304
image dataset
  processing 97
image filtering
  BLUR filter 279
  COUNTOUR filter 280
  EDGES_ENHANCE_MORE filter 280
  filter method, used 279, 280
  FIND_EDGES filter 279
ImageFilter object
  reference documentation 280
  using 279
ImageOps object
  URL 282
image processing operations
  invert operation 281, 282
image processing, with PIL
  filtering 279, 280
  histogram 277
  image, opening 277
  image transformations 282, 283

operations 281
image similarity search 93, 94
image transformations 282, 283
input collection, MapReduce
  filtering 258, 259
insert method, MongoDB 229
integrate method 157
interaction 74
IP address validation 36
IPython
  about 295
  multiprocessing 295
  URL 270
IPython Notebook (NB)
  about 273-275
  blank notebook, starting 273, 274
  data, sharing 296-298
  data visualization 275
  sharing 296

J

JavaScript 53
JavaScript file (.js) 52
JSON
  about 39
  GDF, transforming to 190
JSON file
  parsing, json module used 39, 40
JSON (JavaScript Object Notation) 22, 228

K

Kaggle
  URL 27
kernel functions, SVM 145
Kernel Ridge Regression (KRR) 21, 126-128
knlRidge.pred() method 129
knowledge domain 9

L

language detection 80
LDA 140
learning 80
learn method 127
Linear Discriminant Analysis. *See* LDA

S

Scalable Vector Graphics. *See* **SVG**
scatter_matrix method **291**
scatter plots
  about **64, 65**
  Math.random() function **65**
Scientific Data from University of Muenster
  URL **27**
SciKit 20
SciPy
  about **20, 308**
  installing, on Ubuntu **308**
  installing, on Windows **309**
  running, on Ubuntu **308**
  running, on Windows **309**
  URL **308**
search engines 80
search() function 35
search, Twython
  performing **204, 205**
seasonal influenza (Flu) data
  URL **155**
selectAll function 60
sensors
  QR code (Quick Response Code) **18**
  RFID (Radio-frequency identification) **18**
  using **18**
Sentiment140
  about **217**
  URL **217**
sentiment analysis
  about **200**
  performing, for tweets **221, 222**
sentiment classification
  about **216**
  ANEW **217**
  general process **216**
  text corpus **217**
Series 283
Sharding 228
similarity-based image retrieval
  DTW **94**
  DTW, implementing **97**
  image dataset, processing **97**
  image similarity search **93**
  implementing **93**

  results, analyzing **101-103**
single line chart
  about **67, 68**
  line element **69**
SIR model
  about **156, 157**
  ordinary differential equation, solving with
    SciPy **157, 159**
SIR_model function 157
SIRS model 159, 160
SIRS model simulation
  performing in CA, with D3.js **163-172**
smoothed time series
  predicting **130, 131**
Smoothing Window 123
Social Networks Analysis (SNA) 19, 20, 177
SpamAssassin 82
spam classification 80
spam dataset
  URL **83**
spam text 84
speech recognition 80
SQL 28, 29
SQL databases 29
statistical analysis
  about **183**
  male to female ratio **184, 185**
statistical methods, data scrubbing
  about **34**
  values **34**
statistics 8
statistics function 169
Structured Query Language. *See* **SQL**
support vector machine. *See* **SVM**
SVG 54
SVM
  about **21, 126, 135, 144**
  double spiral problem **145**
  implementing **144**
  implementing, on mlpy **146-150**
  kernel functions **145**

T

text classification
  about **79**
  algorithm **86-88**

Thank you for buying
Practical Data Analysis

About Packt Publishing

Packt, pronounced 'packed', published its first book "*Mastering phpMyAdmin for Effective MySQL Management*" in April 2004 and subsequently continued to specialize in publishing highly focused books on specific technologies and solutions.

Our books and publications share the experiences of your fellow IT professionals in adapting and customizing today's systems, applications, and frameworks. Our solution based books give you the knowledge and power to customize the software and technologies you're using to get the job done. Packt books are more specific and less general than the IT books you have seen in the past. Our unique business model allows us to bring you more focused information, giving you more of what you need to know, and less of what you don't.

Packt is a modern, yet unique publishing company, which focuses on producing quality, cutting-edge books for communities of developers, administrators, and newbies alike. For more information, please visit our website: www.packtpub.com.

About Packt Open Source

In 2010, Packt launched two new brands, Packt Open Source and Packt Enterprise, in order to continue its focus on specialization. This book is part of the Packt Open Source brand, home to books published on software built around Open Source licences, and offering information to anybody from advanced developers to budding web designers. The Open Source brand also runs Packt's Open Source Royalty Scheme, by which Packt gives a royalty to each Open Source project about whose software a book is sold.

Writing for Packt

We welcome all inquiries from people who are interested in authoring. Book proposals should be sent to author@packtpub.com. If your book idea is still at an early stage and you would like to discuss it first before writing a formal book proposal, contact us; one of our commissioning editors will get in touch with you.

We're not just looking for published authors; if you have strong technical skills but no writing experience, our experienced editors can help you develop a writing career, or simply get some additional reward for your expertise.

Building Machine Learning Systems with Python

ISBN: 978-1-78216-140-0   Paperback: 290 pages

Master the art of machine learning with Python and build effective machine learning systems with this intensive hands-on guide

1. Master Machine Learning using a broad set of Python libraries and start building your own Python-based ML systems

2. Covers classification, regression, feature engineering, and much more guided by practical examples

3. A scenario-based tutorial to get into the right mind-set of a machine learner (data exploration) and successfully implement this in your new or existing projects

Clojure Data Analysis Cookbook

ISBN: 978-1-78216-264-3   Paperback: 342 pages

Over 110 recipes to help you dive into the world of practical data analysis using Clojure

1. Get a handle on the torrent of data the modern Internet has created

2. Recipes for every stage from collection to analysis

3. A practical approach to analyzing data to help you make informed decisions

Please check **www.PacktPub.com** for information on our titles

Hadoop Operations and Cluster Management Cookbook

ISBN: 978-1-78216-516-3   Paperback: 368 pages

Over 60 recipes showing you how to design, configure, manage, monitor, and tune a Hadoop cluster

1. Hands-on recipes to configure a Hadoop cluster from bare metal hardware nodes

2. Practical and in depth explanation of cluster management commands

3. Easy-to-understand recipes for securing and monitoring a Hadoop cluster, and design considerations

4. Recipes showing you how to tune the performance of a Hadoop cluster

KNIME Essentials

ISBN: 978-1-84969-921-1   Paperback: 130 pages

Perform accurate data analysis using the power of KNIME

1. Learn the essentials of KNIME, from importing data to data visualization and reporting

2. Utilize a wide range of data processing solutions

3. Visualize your final datasets using KNIME's powerful data visualization options

Please check **www.PacktPub.com** for information on our titles

www.ingramcontent.com/pod-product-compliance
Lightning Source LLC
Chambersburg PA
CBHW062053050326
40690CB00016B/3076